Design of Low-Noise Amplifiers for Ultra-Wideband Communications

Roberto Díaz Ortega
Sunil Lalchand Khemchandani
Hugo García Vázquez
Francisco Javier del Pino Suárez

Mc
Graw
Hill
Education

New York   Chicago   San Francisco
Athens   London   Madrid
Mexico City   Milan   New Delhi
Singapore   Sydney   Toronto

Cataloging-in-Publication Data is on file with the Library of Congress.

McGraw-Hill Education books are available at special quantity discounts to use as premiums and sales promotions, or for use in corporate training programs. To contact a representative please visit the Contact Us page at www.mhprofessional.com.

Design of Low-Noise Amplifiers for Ultra-Wideband Communications

1 2 3 4 5 6 7 8 9 0 DOC/DOC 1 9 8 7 6 5 4 3

ISBN 978-0-07-182312-8
MHID 0-07-182312-3

The pages within this book were printed on acid-free paper.

Sponsoring Editor
Michael McCabe

Proofreader
Aptara, Inc.

Acquisitions Coordinator
Bridget Thoreson

Production Supervisor
Pamela A. Pelton

Editorial Supervisor
David E. Fogarty

Composition
Aptara, Inc.

Project Manager
Indu Jawad, Aptara, Inc.

Art Director, Cover
Jeff Weeks

Copy Editor
Gail Naron Chalew

Preface

In recent years, wireless personal area networks (WPANs) have become popular, replacing cables and enabling new consumer applications. Such systems are now dominated by standards such as Bluetooth and Zigbee, which operate in the 2.4 GHz ISM band. In order to improve the data rate to several hundreds of Mb/s with a low power transmission, ultra-wideband (UWB) communications have been proposed.

In a classical zero-IF receiver architecture, one of the most challenging components is the low-noise amplifier (LNA). This circuit must have a precise amplification over a wide range of frequencies with a wideband input matching and a low noise contribution. Due to these strict requirements, low-noise amplifiers are usually composed of a large numbers of inductors having a high power consumption.

In mobile applications the power consumption is directly related to battery life. On the other hand, the area consumption is related to fabrication costs. In order to get commercial solutions, it is fundamental to obtain a low-cost implementation with a low power consumption.

In this book, different alternatives to implement power and area efficient integrated low-noise amplifiers for UWB communications based on the ECMA-368/ISO/IEC 26907 standard are presented. Design methodologies for distributed amplifiers, feedback amplifiers, inductor structures with reduced area, and inductorless design techniques are addressed.

The system design is presented in Chapter 1, which describes the main requirements of ECMA-368/ISO/IEC 26907. With these requirements, a reference system is obtained and the specifications of the LNA are extracted. Chapter 2 is devoted to the most classical wideband amplifier architecture, the distributed amplifier. After this first approach and with the objective of solving distributed amplifier's drawbacks, in Chapter 3 different implementations of wideband LNAs are presented. To continue reducing area and power consumption, in Chapter 4 feedback techniques and some inductors' structures suited for those topologies are explored. Chapter 5 explores inductorless techniques to reduce the area of low-noise amplifiers.

This book will enable the engineer to design wideband LNAs using standard CMOS technologies.

Nomenclature

ADC	Analog to digital converter
ADS	Advanced Design System
AGC	Automatic gain control
AMS	Austria Mikro Systeme
AWGN	Additive white gaussian noise
BER	Bit error rate
BiCMOS	Bipolar CMOS
BW	Bandwidth
CG	Common gate
CMOS	Complementary metal-oxide semiconductor
DA	Distributed amplifier
DLNA	Digital Living Network Alliance
ECMA	European Computer Manufacturers Association
EM	Electromagnetic simulation
FCC	Federal Communications Commission
FOM	Figure of merit
IEC	International Electrotechnical Commission
IEEE	Institute of Electrical and Electronics Engineers
IF	Intermediate frequency
IIP3	Input referred third-order intercept point
IP	Internet Protocol
ISO	International Organization for Standardization
LNA	Low-noise amplifier
LO	Local oscillator
MAC	Media Access Controller
MB-OFDM	MultiBand OFDM
MBOA	MultiBand OFDM Alliance
MLS	Multi-level structure
MOS	Metal-oxide-semiconductor field-effect transistor

NF   Noise figure
OFDM   Orthogonal frequency-division multiplexing
PCB   Printed circuit board
PER   Packet error rate
PHY   Physical layer
PSDU   Physical layer service data unit
Q   Quality factor
RF   Radio frequency
SAW   Surface acoustic wave
SLS   Single-level structure
SNR   Signal-to-noise ratio
UWB   Ultra-wideband
VCO   Voltage-controlled oscillator
WiFi   Wireless fidelity
WLAN   Wireless local area network
WPAN   Wireless personal area network

CHAPTER **1**

Ultra-Wideband Overview and System Approach

1.1 Introduction

As a starting point for this book, this chapter discusses ECMA-368/ISO/IEC 26907 standard. After a brief summary of its history and main specifications, the chapter develops an analysis of the receiver, which takes into account the restrictions and specifications imposed by the standard. The obtained receiver specifications are taken as a reference point for the circuits designed in the rest of the book.

1.2 History of Ultra-Wideband Communications

UWB technology originated around 1962 when it was referred to impulse radio or baseband carrier-free communications. However, the term "ultra-wideband" was first used in 1989 in a patent document by the U.S. Defense Department.

In 2002, the Federal Communications Commission (FCC), with the inform 02-48, allocated an unlicensed radio spectrum of 3.1–10.6 GHz. To define a device as an UWB device, its channels have to occupy a band greater than 20 percent of the center frequency or a minimum channel bandwidth of 500 MHz.

After the first attempt of standardization by the FCC, the MultiBand OFDM Alliance (MBOA) was established in 2003. It is dedicated to promoting the global standard for ubiquitous UWB wireless solutions. The MBOA created a complete specification for a physical layer (PHY) and a Media Access Controller layer (MAC).

Wireless USB	**Wireless IEEE-1394**	**DLNA**	**WiMedia Profile**
		TCP/IP Adaptation Layer	

WiMedia Covergence Platform

MBOA specifications

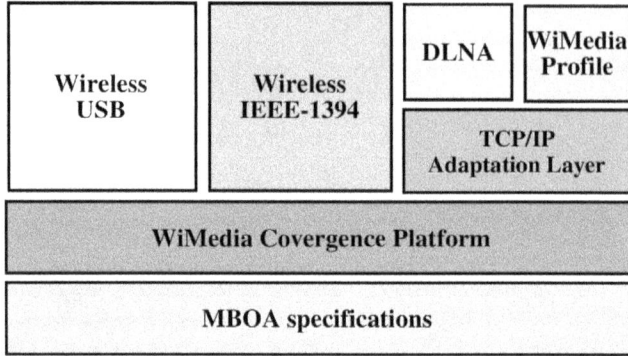

FIGURE 1.1 WiMedia layers stack.

In parallel to the MBOA, the IEEE 802.15.3a task group was created to study the possibility of using the new FCC spectrum specifications in wireless local area networks (WLANs) and personal area networks (WPANs).

Outside the IEEE 802.15.3a, different companies formalized their relationships to provide a legal context. From this formalization, in 2004 the WiMedia Alliance was born to promote wireless connectivity and interoperability among multimedia devices. The objective of WiMedia is to develop a common abstraction platform as shown in Fig. 1.1, which enables multiple applications to run over one common radio. The WiMedia radio platform is based technically on MBOA specifications. The combination of MBOA and this convergence platform allows the implementation of wireless versions of USB, IEEE 1394, DLNA, and other IP-based application protocols.

On January 2006, after three years of a intensive process, the IEEE 802.15.3a was abandoned without conclusion. Without the support of the IEEE, the WiMedia Alliance had to seek an alternative means of standardizing UWB communications. After much hard work, in 2007 the first version of standard ECMA-368/ISO/IEC 26907 was approved that regulates the UWB communications at the Physical and Media Access Controller layers.

1.3 ECMA-368/ISO/IEC 26907 Receiver Specifications

1.3.1 Operating Frequency Band

The physical layer operates in a frequency range from 3.1 to 10.6 GHz. The relationship between the center frequency (fc) and the channel number (BAND_ID number or n_b) is given by Eq. 1.1.

$$f_c(n_b) = 2904 + 528 \cdot n_b \; MHz \qquad where \; n_b = 1, ..., 14 \qquad (1.1)$$

This definition provides a unique numbering system for all channels that have a spacing of 528 MHz. As defined in Fig. 1.2, six band groups are defined. Band groups 1 to 4 consist of three bands each, spanning the band 1 to 12. Band group 5 contains the two bands 13 and 14. Band group 6 contains the bands 9, 10, and 11. The allocation band is summarized in Table 1.1. In spite

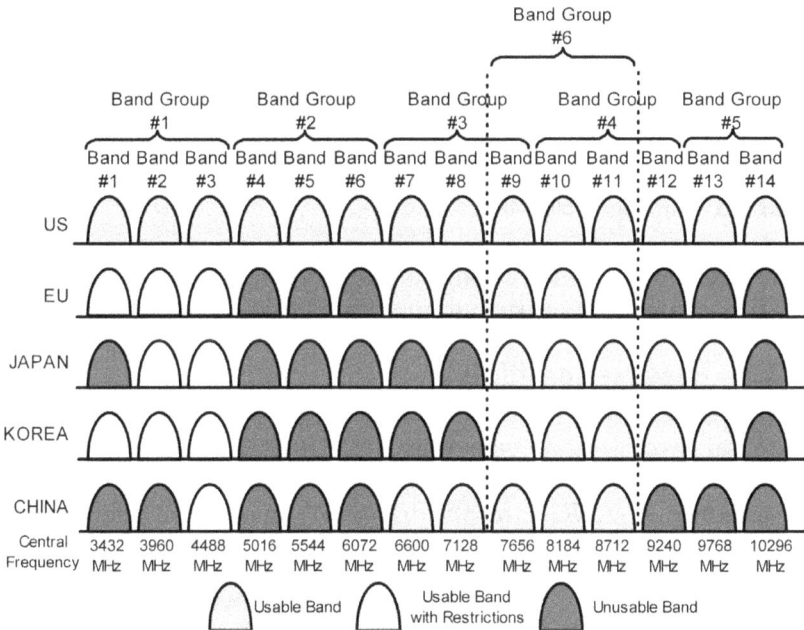

FIGURE 1.2 UWB operating band limitations.

Band group	Band ID (n_b)	Lower frequency (MHz)	Center frequency (MHz)	Upper frequency (MHz)
1	1	3168	3432	3696
	2	3696	3960	4224
	3	4224	4488	4752
2	4	4752	5016	5280
	5	5280	5544	5808
	6	5808	6072	6336
3	7	6336	6600	6864
	8	6864	7128	7392
	9	7392	7656	7920
4	10	7920	8184	8448
	11	8448	8712	8976
	12	8976	9240	9504
5	13	9504	9768	10032
	14	10032	10296	10560
6	9	7392	7656	7920
	10	7920	8184	8448
	11	8448	8712	8976

TABLE **1.1**   Band group allocation.

of this recommendation about the allocation bands, each country can create more restrictive rules about frequency band use.

1.3.2   Receiver Sensitivity

For a packet error rate (PER) of less than 8 percent with a physical layer service data unit (PSDU) of 1024 octets, the minimum receiver sensitivity numbers with an additive white gaussian noise (AWGN) for the different data rates are listed in Table 1.2.

1.4 Receiver System Design

Due to the huge channel bandwidth, a direct conversion receiver architecture has been chosen. Figure 1.3 shows the direct

Data rate (Mb/s)	Sensitivity (dBm)
53.3	−80.8
80	−78.9
106.6	−77.8
160	−75.9
200	−74.5
320	−72.8
400	−71.5
480	−70.4

TABLE 1.2   Sensitivity versus data rate.

conversion receiver diagram block, where the local oscillator (LO) frequency is equal to the input carrier frequency. Note that channel selection requires only a low-pass filter with relative sharp cut-off characteristics.

This architecture has several limitations. First, in a direct conversion topology, the down converted band extends to zero frequency. As a result, offset voltages can corrupt the signal and saturate the following stages. This issue is also related to the LO leakage because the LO radiation could appear as a DC voltage at the receiver output. Second, as shown in Fig. 1.3, phase and frequency modulation requires shifting either radio frequency (RF) or LO signal output by 90°. This shifting generally introduces errors and noise. Due to this error I/Q mismatches could appear, thereby raising the bit error rate. Third, in baseband, the even-order harmonics could be into the desired channel.

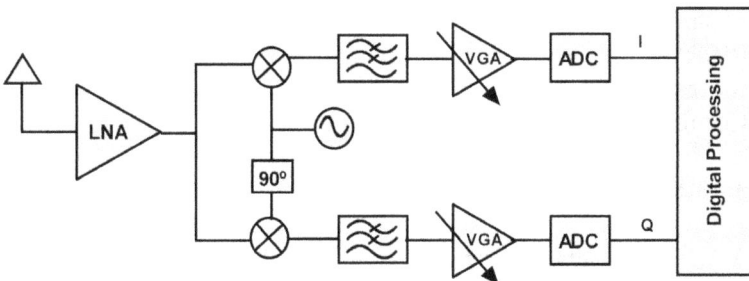

FIGURE 1.3   Direct conversion architecture.

Fourth, because the desired channel is translated directly to baseband, the flicker noise could affect the signal.

On the contrary, the simplicity of the direct conversion architecture offers two important advantages. First, the problem of the image frequency does not appears. As a result, no image filter is required. Second, the intermediate frequency (FI) surface acoustic wave (SAW) filter and other downconversion stages, used for instance in heterodyne receivers, are replaced with low-pass filters and baseband amplifiers, so this architecture is more suitable for a monolithic integration with a relatively low area and low power consumption.

1.4.1   Noise Figure

As shown in previous section, the ECMA-368/ISO/IEC 26907 specifies a sensitivity of −70.4 dBm (S_{in}) for the highest data rate of 480 Mb/s (R) and a sensitivity of −80.8 dBm for the lowest data rate (53.3 Mb/s) with a channel bandwidth (BW) of 528 MHz. The noise figure is defined as the degradation of the signal to noise ratio as it is shown in Eq. 1.2:

$$NF = SNR_{in} - SNR_{out} \qquad (1.2)$$

where,

$$SNR_{in} = S_{in} - (174 + 10 \cdot log(BW)) \qquad (1.3)$$

$$SNR_{out} = \left(\frac{E_b}{N_0}\right)_{dB} + \left(\frac{R}{BW}\right)_{dB} \qquad (1.4)$$

To obtain the value of energy per bit to noise power spectral density ratio (E_b/N_0), the standard defines a quaternary phase shift keying (QPSK) modulation for each subcarrier and a PER of 8 percent for a 1024 byte packet. With this PER a bit error rate (BER) of 10^{-5} is obtained. Finally from Fig. 1.4 can be obtained the estimated value of E_b/N_0 (10 dB) for the calculated BER. Combining Eqs. 1.3 and 1.4, Table 1.3 shows the maximum and minimum noise figure values for different values of sensitivity and data rate.

1.4.2   Channel Filter and ADC Specifications

In the frequency band occupied by ECMA-368/ISO/IEC 26907, the most harmful interference is caused by WiFi (IEEE 802.11a)

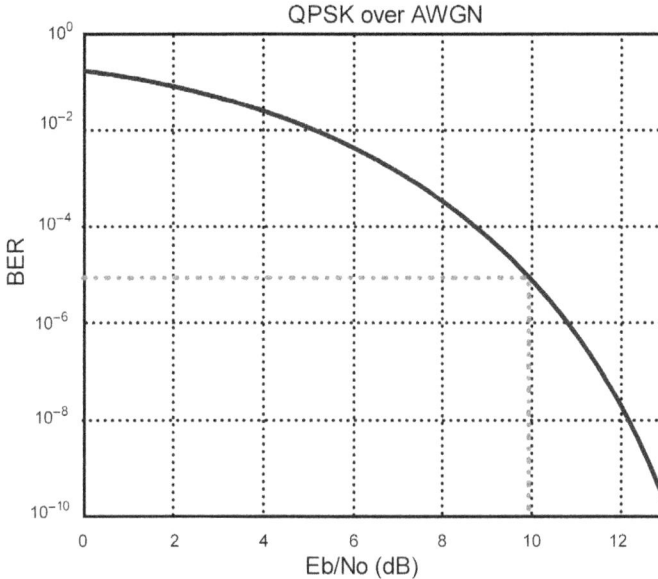

QPSK over AWGN

FIGURE **1.4** BER estimation.

channels in the 5 to 6 GHz band. According to standard specifications, the adjacent channels must be attenuated 36 dB (Fig. 1.5).

The channel filtering is also defined as a tradeoff between the analog to digital converter (ADC) and the channel filter. The filter specifications could be relaxed by increasing the ADC dynamic range and removing the interference through digital processing.

The dynamic range of the ADC (DR_{ADC}) is defined by Eq. 1.5,[1] where A_{filter} is the filter attenuation. It is important to avoid saturating the ADC at the maximum input power according to Eq. 1.5 and taking into account the worst case: at minimum input power (P_{inMin}) of −80.8 dBm and a maximum

Data rate (MB/s)	Sensitivity (dBm)	Noise figure (dB)
480	−70.4	7.32
53.3	−80.8	18.9

TABLE **1.3** Receiver noise figure values.

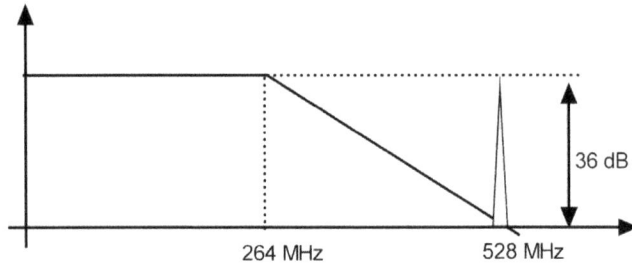

FIGURE 1.5 Filter requirements.

input power (P_{inMax}) of -30 dBm (WiFi interference), the ADC dynamic range is easily determined from Eq. 1.5.

$$DR_{ADC} = \underbrace{(P_{inMax} - A_{Filter})}_{P_{max}} - \underbrace{(P_{inMin} - 15\,dB)}_{quantization\ noise} \qquad (1.5)$$

Once the dynamic range of the ADC has been obtained, the ADC number of bits could be determined using Eq. 1.6:[1]

$$DR_{ADC} = 6.02 \cdot N + 1.76\,dB \qquad (1.6)$$

With different filter specifications, ADC dynamic ranges, and ADC number of bits, different configurations can be obtained. Table 1.4 shows different combinations for the filter and ADC specifications.

According to the standard specifications, the maximum input power is -41 dBm for a desired in-band signal. In the worst case, the maximum dynamic range is 39.8 dB. In consequence, in Table 1.4 the third configuration must be rejected because in this configuration the ADC will be saturated with the input power worst case.

Filter roll-off (dB/oct)	Filter order	ADC dynamic range (dB)	ADC bit numbers
12	2°	53.8	≥ 9
24	4°	41.8	≥ 7
36	6°	29.8	≥ 5

TABLE 1.4 Channel filter and ADC specifications.

1.4.3 ADC and Frontend Gain Specifications

The quantization noise is given by:[1]

$$N_Q = \frac{\left(\dfrac{V_{FS}}{2^N}\right)^2}{12 \cdot \left(\dfrac{2 \cdot f_s}{f_{bW}}\right) \cdot R_o} = \frac{\left(\dfrac{V_{FS}}{2^N}\right)^2}{12 \cdot p \cdot \rho_o} \qquad (1.7)$$

Expressed in dBm with $R_o = 50 \ \Omega$, the quantization noise results in:

$$N_Q = -6.02 \cdot N + 20 \cdot log(V_{FS}) - 10 \cdot log(\rho) + 2.2 \ dBm \qquad (1.8)$$

where, V_{FS} is the ADC full range voltage input, N is the ADC number of bits, and ρ is the oversampling factor expressed as $\rho = 2 \cdot f_s / BW$. Considering a V_{FS} of 2V the quantization noise is given by:

$$N_Q = -6.02 \cdot N - 10 \cdot log(p) + 8.22 \qquad (1.9)$$

On the other hand, the output thermal noise is given by the following expression:

$$N_T = -174 + 10 \cdot log(BW) + G_{max} + NF = G_{max} - 79.77 \qquad (1.10)$$

Combining Eqs 1.9 and 1.10 and taking into account that $N_T = N_Q + 15$, the minimum total gain of the receiver could be expressed as:

$$G_{min} = -6.02 \cdot N - 10 \cdot log(p) + 102.99 \qquad (1.11)$$

In Table 1.5, the gain values for different oversampling factors and ADC number of bits are shown.

ADC bits	Oversampling factor (p)	G_{min} (dB)
7	1	60.86
	2	57.86
9	1	48.81
	2	45.81

TABLE 1.5 Minimum receiver gain specifications.

1.4.4 Automatic Gain Control

In the previous section, the minimum gain value was determined to obtain a quantization noise level 15 dB less than the thermal noise. However, the calculated gain values could saturate the ADC input. The maximum power level at the ADC input is given by:

$$P_{ADC} = \frac{\left(\frac{V_{FS}}{2}\right)^2}{2 * R_0} = 10 \; mW \Rightarrow 10 \; dBm \tag{1.12}$$

In consequence, to avoid saturating the ADC, the system should satisfy the following expression:

$$S_{Imax} + G_{min} \leq 10 \; dBm \tag{1.13}$$

In the worst case, the maximum desired input power (S_{Imax}) is −41 dBm. Under this condition, the maximum permitted gain for the entire reception chain should be less than 51 dB because with this gain value, the power level at ADC input is 10 dBm. As observed in Table 1.5, the ADC number of bits should be fixed at least to 9 bits, because with this number of bits, the minimum system gain is less than the maximum permitted gain at the ADC input.

On the other hand, to obtain the maximum receiver gain, the ADC dynamic range should be considered. From Table 1.4, an ADC of 9 bits should have a dynamic range of 53.8 dB. The ADC dynamic range is defined by the following expression:

$$\Delta AGC = S_{Imax} - S_{Imin} \tag{1.14}$$

Table 1.5 shows the receiver gain specifications; as mentioned earlier, the maximum power level at the ADC input is 10 dBm so, from Eq. 1.14, the minimum ADC input signal level is: −43.8 dBm. On the other hand, for the weak signal and the minimum gain established in 48.81 dB, so in this condition the power level at the ADC input is −32 dBm.

Concluding from the previous results, the automatic gain control (AGC) is not needed because in the minimum and maximum input power level condition, the ADC input is not saturated.

1.4.5   Linearity Requirements

The interference scenario is dominated by IEEE 802.11a. In a typical case, an IEEE 802.11a channel at a distance of 0.2 m could reach a power level of −31.9 dBm. This interference should coexist with a desired ECMA-368/ISO/IEC 26907 signal with a power level of −80.8 dBm. From this interference scenario, the linearity is defined by the following expression:

$$IIP3 = S_{desired} + \frac{3}{2} \cdot (S_{interference} - S_{desired}) \Rightarrow IIP3 \geq -8.65 \, dBm$$

(1.15)

where $S_{desired}$ is the desired signal power and $S_{interference}$ is interference signal power.

1.4.6   Synthesizer Requirements

Because the radio has to cover the six bands defined in the ECMA-368/ISO/IEC 26907 standard and a zero-IF architecture is proposed, the synthesizer should provide the center frequencies of the bands shown in Table 1.1.

In the MBOA proposal, frequency hopping between sub-bands occurs once every symbol period of 312.5 ns. This period contains a 60.6 ns suffix, which is followed by a 9.5 ns guard interval. The frequency generator used to drive the switching core of both the down-conversion mixer in the receive path and the up-conversion mixer in the transmit path needs to switch within this 9.5 ns to accomplish the frequency hopping.

The demands on the purity of the generated carriers are also very stringent due to the presence of strong interferer signals. For example, for Mode 1 operation all spurious tones in the 5 GHz range must be below 50 dBc to avoid down-conversion of strong out-of-band WLAN interferers into the wanted bands. For the same reason, the spurious tones in the 2 GHz range should be below 45 dBc to allow coexistence with the systems operating in the 2.4 GHz ISM band, such as IEEE 802.11b/g and Bluetooth.

Finally, to ensure that the system SNR will not be degraded by more than 0.1 dB due to the LO generation, the voltage-control oscillator (VCO) phase noise specification is set to 100 dBc/Hz at 1 MHz offset, and the overall integrated phase noise should not exceed 3.5 degrees rms.[2,3]

14

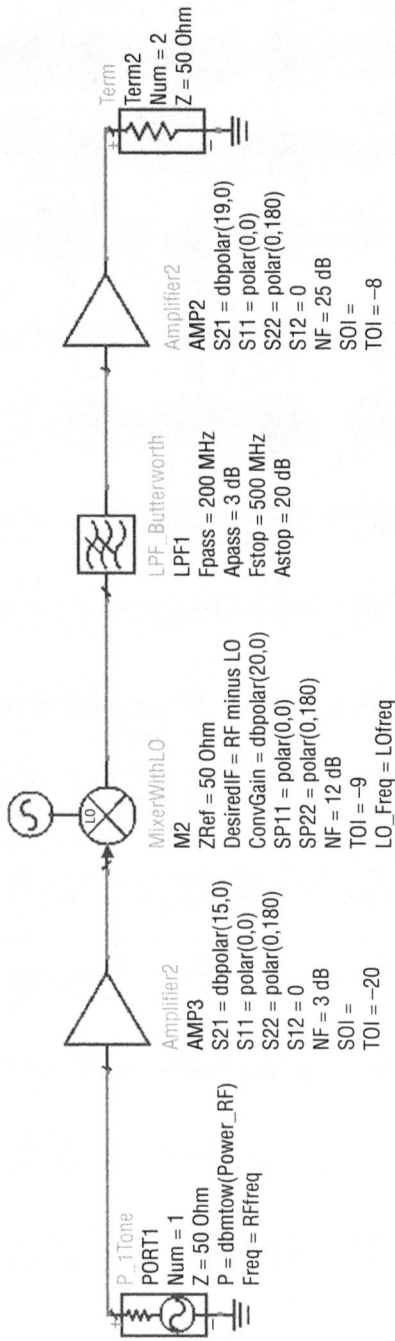

P_1Tone
PORT1
Num = 1
Z = 50 Ohm
P = dbmtow(Power_RF)
Freq = RFfreq

Amplifier2
AMP3
S21 = dbpolar(15,0)
S11 = polar(0,0)
S22 = polar(0,180)
S12 = 0
NF = 3 dB
SOI =
TOI = −20

MixerWithLO
M2
ZRef = 50 Ohm
DesiredIF = RF minus LO
ConvGain = dbpolar(20,0)
SP11 = polar(0,0)
SP22 = polar(0,180)
NF = 12 dB
TOI = −9
LO_Freq = LOfreq

LPF_Butterworth
LPF1
Fpass = 200 MHz
Apass = 3 dB
Fstop = 500 MHz
Astop = 20 dB

Amplifier2
AMP2
S21 = dbpolar(19,0)
S11 = polar(0,0)
S22 = polar(0,180)
S12 = 0
NF = 25 dB
SOI =
TOI = −8

Term
Term2
Num = 2
Z = 50 Ohm

FIGURE 1.6 System simulations schematic.

Component	Gain (dB)	Noise figure (dB)	IIP3 (dBm)
LNA	15	3	−20
Mixer	20	12	−9
Baseband filter	−3	3	−
Baseband amplifier	19	25	−8

TABLE **1.6**   Receiver block specifications.

1.4.7   Budget Simulations

To obtain the receiver block specifications, the simulation tool Advanced Design System (ADS) has been used. The budget simulation checks the receiver chain performance to verify that the specifications of each block of the receiver allow to the entire reception chain to fulfill the specifications stated in the previous sections.

As starting point, a first assumption for each block parameter is set, with the help of the state of the art and the design group experience. The final value of the specifications is set using the simulation tool ADS with the budget analysis and a iterative simulation process. Figure 1.6 shows the simulation schematic for the budget analysis. It can be observed how the entire receiver chain follows the schematic shown in Fig. 1.3.

Table 1.6 summarizes the final specifications of each receiver component, obtained from simulations. As observed in Table 1.7, the budget simulation results are compliant with the global receiver requirements.

Figure 1.7 shows the SNR variation through the receiver. Obviously, at the receiver input, the SNR is the highest;

Receiver parameter	Specification	Budget simulation
Sensitive (dBm)	−80.8	−85
Noise figure (dB)	7.32	7.27
Gain (dB)	48.81	50.9
Maximum input level (dBm)	−41	−35
IIP3 (dBm)	−8.65	−8.15

TABLE **1.7**   Budget simulation results.

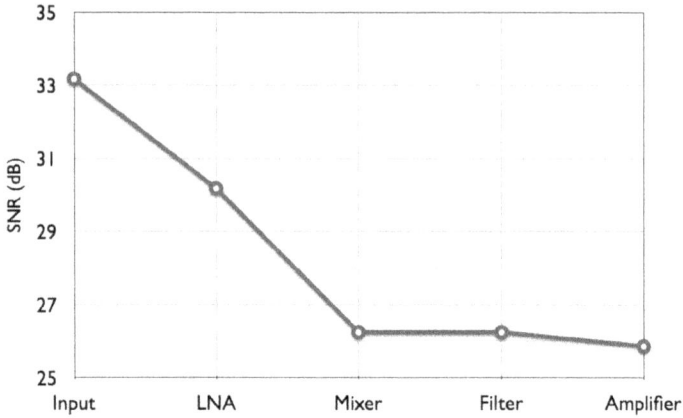

FIGURE 1.7 SNR budget simulation.

however, as the signal moves through the receiver chain, it is corrupted by noise and in consequence the SNR drops. The difference between the SNR at the input and output of the receiver is the noise figure. In this case, there is a different between input and output of 7.2 dB.

Figure 1.8 shows the gain contribution of each block to the entire receiver. As observed, the gain specification is divided equally between the LNA, the mixer, and the baseband amplifier. This situation provides a relaxed scenario for circuit designers because if one block does not reach the proposed gain,

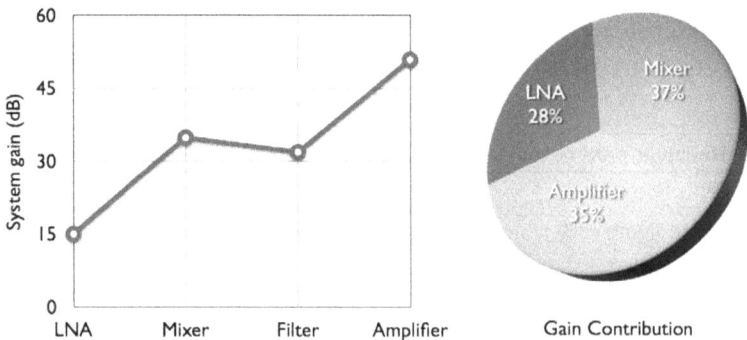

FIGURE 1.8 Gain budget simulation.

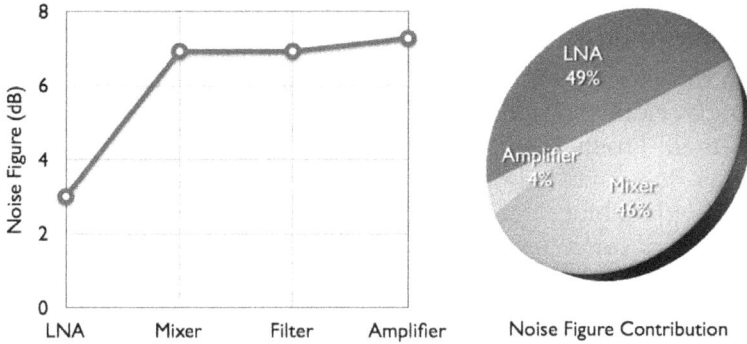

FIGURE 1.9   Noise figure budget simulation.

this problem could be resolved by increasing the gain in a subsequent block.

As it observed in Fig. 1.9, the noise figure contribution depends mainly on the LNA and mixer, so the design of those circuits will be fundamental to satisfying the noise figure specifications.

Finally, Fig. 1.10 shows the contribution of each individual block to the receiver linearity. In this case, the main contributions are established by the mixer and the baseband amplifier. In this situation the circuit designers will have to focus their effort in obtaining the maximum linearity in both circuits.

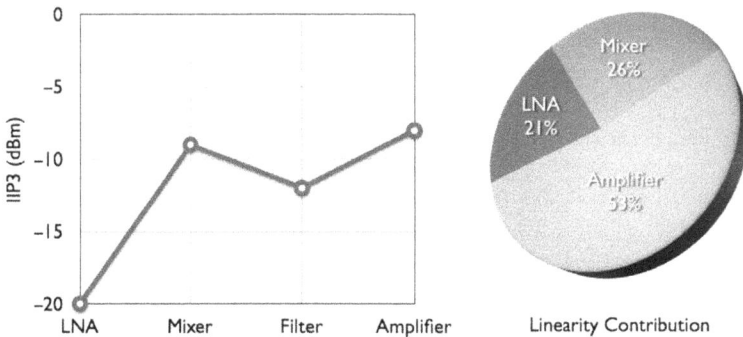

FIGURE 1.10   Linearity budget simulation.

1.5 Conclusions

In this chapter, the history and the main characteristics of the ultra-wideband standard ECMA-368/ISO/IEC 26907 were presented. From these specifications, a receiver system analysis was done and validated through simulations, obtaining the individual block specifications.

As a summary, Table 1.8 shows the low-noise amplifier specifications. These specifications will be taken as reference points to develop low-noise amplifiers in the following chapters.

The next chapter is devoted to the distributed amplifier, one of the most classical structures used to develop wideband LNAs.

Parameter	Value
Gain (dB)	15
Noise Figure (dB)	3
IIP3 (dBm)	−20
Power Consumption (mW)	minimum
Area Consumption (mm^2)	minimum

TABLE 1.8   Low-noise amplifier specifications.

CHAPTER 2

Distributed Amplifiers

2.1 Introduction

The design of low-noise amplifiers (LNAs) for ultra-wideband communications has to solve a big challenge: obtaining a huge bandwidth. Using a distributed amplifier is the first approach to obtain an LNA for UWB systems. With this structure a high bandwidth with a relative low noise and a moderate gain can be obtained.

2.2 Theoretical Approach

The frequency response of a metal-oxide semiconductor field-effect transistor (MOSFET) device degrades due to the pole formed by the input/output capacitance of the transistor and the resistance it sees. The MOSFET's transconductance rapidly falls with frequency, and any attempt to increase the transconductance by increasing the size of the device will also increase its input/output capacitance. Thus, while low-frequency gain has been increased, the gain-bandwidth product remains about the same. Distributed amplification (DA) was proposed to overcome this limitation.

A distributed amplifier employs a topology in which the gain stages are connected such that their capacitances are separated, yet the output currents still combine in an additive fashion (Fig. 2.1). Series inductive elements are used to separate capacitances at the inputs and outputs of adjacent gain stages. The resulting topology, given by the interlaying series inductors and shunt capacitances, forms a lumped-parameter artificial transmission line. The additive nature of the gain dictates a relatively low gain; however, the distributed nature of the capacitance allows the amplifier to achieve very wide bandwidths.

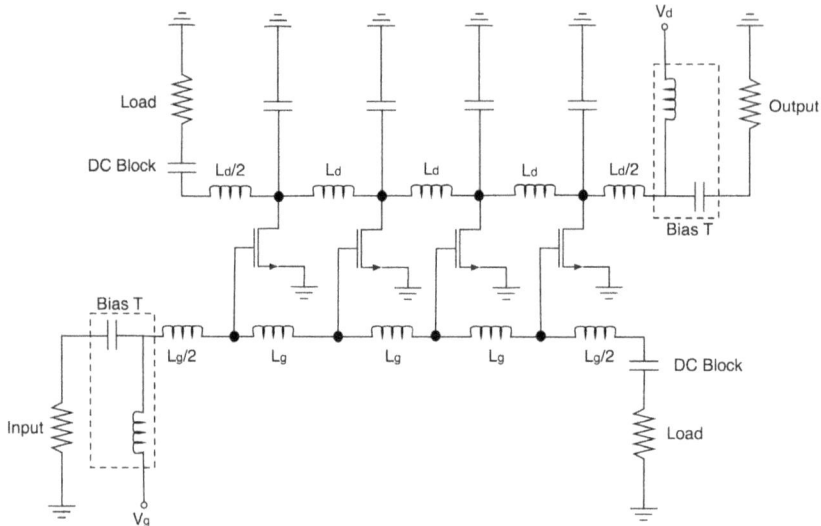

FIGURE 2.1   Distributed amplifier schematic.

Distributed amplification overcomes the gain-bandwidth limitation, absorbing the MOSFET input/output capacitance as
part of the lumped elements of the artificial transmission line,
formed with the series inductance that connects adjacent drains
and gates. The characteristic impedance (Z_0) and cut-off frequency (f_c) of lossless transmission line are given at a first
approximation by:

$$Z_0 = \sqrt{\frac{L_{TL}}{C_{TL}}} \tag{2.1}$$

$$f_c = \frac{1}{\pi \sqrt{L_{TL}C_{TL}}} \tag{2.2}$$

where subindex TL accounts for the drain and gate transmission lines. Since Z_0 and f_c of both the drain and gate lines are the
same, their capacitances and inductances should be the same.
Because the drain-to-bulk capacitance C_{db} of a MOSFET is usually less than its gate-to-source capacitance C_{gs}, a capacitor C_d
is added to the drain to make the capacitances equal.

$$L_g = L_d \tag{2.3}$$

$$C_{gs} = C_{db} + C_d \tag{2.4}$$

As the amplified signals at each stage travels toward the load, the signal gets attenuated due to non-zero losses associated with the transmission lines. Finite inductors quality factor (Q) are the primary source of losses in the gate line. Losses in the drain line can be attributed to lossy inductors L_d and the drain-to-source resistance (r_{ds}). The gain of the DA can be expressed as:[4]

$$A = -g_m \frac{Z_0}{2\sqrt{1 - \left(\frac{f}{f_c}\right)^2}} \cdot \frac{e^{-N\frac{(A_g + A_d)}{2}} \cdot sinh\left(N\frac{A_d - A_g}{2}\right)}{sinh\left(N\frac{A_d - A_g}{2}\right)}$$

(2.5)

where A_d and A_g are the attenuation of the drain and gate lines, g_m is the MOSFET transconductance, and N is the total number of stages. This equation assumes the following:

- Unilateral MOSFET model (ignores C_{gd}).
- Image impedance matched terminations.
- Equal gate and drain-phase velocities.

The optimum number of stages that maximizes the gain is a function of gate and drain-line attenuation. Those attenuations are complex functions and depend on the specific MOSFET parameters and also on the operating and cut-off frequencies. As the signal propagates along the gate line toward the termination, less signal is available for each MOSFET because of attenuation and, as a consequence, the overall gain degrades with further increase in the number of stages. The number of stages for this work is chosen as four.

Knowing the gain, number of stages, and drain-line inductance and capacitance, the required g_m can be found from the low frequency gain using Eq. 2.5:

$$g_m = \frac{2 \cdot A}{N} \sqrt{\frac{C_d}{L_d}} \Rightarrow g_m = \frac{2 \cdot A}{N \cdot Z_0}$$

(2.6)

Then, the transistor width-length ratio can be derived from

$$\frac{W}{L} = \frac{g_m}{\mu_n C_{ox}(V_{gs} - V_T)}$$

(2.7)

Component	Value
$L_g = L_d$	1.465 nH
$L_g/2 = L_d/2$	1.15 nH
C_d	586 fF
L (transistor length)	0.425 μm
W (transistor width)	4.42 μm

TABLE 2.1   Calculated components values.

where

- W is transistor gate width.
- L is transistor gate length.
- n is electron mobility.
- C_{ox} is gate oxide capacitance per unit area.
- V_T is threshold voltage.
- V_{gs} is gate-source voltage.

Finally, the device length and width can be found by combining Eq. 2.7 with the following expression:

$$W \cdot L = \frac{C_g}{Cox} \tag{2.8}$$

Taking into consideration the previous equations and a four stage structure, a DA with a gain of 10 dB and a cutoff frequency of 11 GHz has been designed. Table 2.1 shows the calculated component values.

An important conclusion can be drawn from the previous analysis: this kind of circuit is composed by of a considerable number of inductors. As noted in the next section, inductors occupy a large amount of layout area, and as a consequence, it is very important to study the effect of inductors and their distribution over the circuit area.

2.3 Area Optimization

2.3.1 Compact Design

The easiest method for reducing the area of a circuit is to compact its layout, which, in the case of a DA, suggests locating

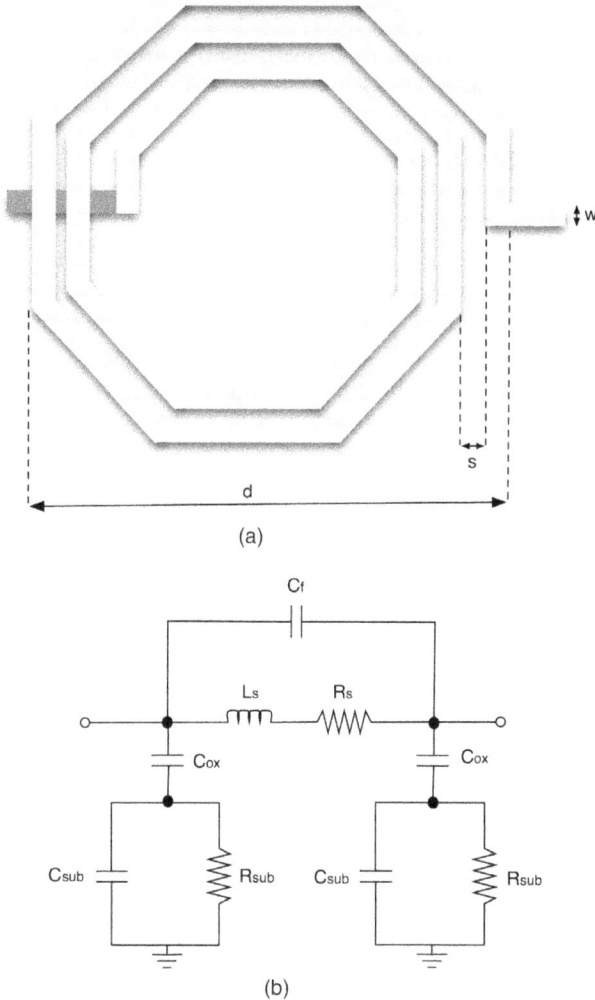

FIGURE 2.2 (a) Layout and design parameters for an on-chip spiral inductor. (b) Simplified lumped-component inductor model.

the inductors as close to each other. Fig. 2.2 (a) shows the typical layout of an on-chip spiral inductor. The design parameters of such a structure are the outer diameter d, the metal width w, the spacing between the wiring metal s, and the number of turns n. The standard lumped-element model associated to this structure is shown in Fig. 2.2 (b).[5] In this model, L_S and R_S represent the series inductance and resistance respectively, C_F is the fringing capacitance between the metal traces and the over-lap capacitance between the spiral inductor and the underpass metal, C_{OX} accounts for the spiral-to-substrate capacitance, and R_{SUB} and C_{SUB} model the behavior of leakage currents across the oxide and the substrate (bulk) and additional capacitive effects related to the substrate.

In complementary metal-oxide semiconductor (CMOS) technologies, on-chip inductors suffer from three main loss mechanisms, namely the ohmic, capacitive, and inductive losses. Ohmic losses result from the current flowing through the resistance of the metal tracks. Those losses can be reduced using a wider metal line, but doing so also increases the capacitive losses (in particular the metal-to-substrate capacitance), caus-ing a decrease in the overall quality factor and self-resonance frequency (f_{SR}). The displacement currents conducted by the metal-to-substrate capacitance and the eddy current generated by the magnetic flux penetrating into de substrate result in ca-pacitive and inductive losses respectively. In the design of DA other undesired effects also appear such as the mutual cou-pling between inductors. This fact worsens if, as stated earlier, inductors are situated close to each other in order to achieve a compact design.

Figure 2.3 (a) shows the electrical model of two series con-nected on-chip inductors, where each inductor has been mod-elled with the simplified lumped-element model shown previ-ously. If the frequency range of interest is limited to few GHz, this model can be simplified by neglecting substrate and metal capacitance. Other elements, such R_S and R_{SUB}, can also be of minor importance for the coupling estimation. So, stripped to its essential, the model is reduced to its inductive elements, resulting in the simplified circuit shown in Fig. 2.3 (b).

This simplified model is composed of two series connected inductors and a mutual coupling between them. Depending on the coiling direction, the total inductance (L_T) can be calculated using Eq. 2.9 or 2.10. Thus, if both inductors are coiled following the same direction, the mutual coupling is positive, and if the spirals are coiled in opposite directions the coupling is negative.

$$L_{T1} = L_{S1} + L_{S2} + 2 \cdot m_{12} \qquad (2.9)$$

$$L_{T2} = L_{S1} + L_{S2} - 2 \cdot m_{12} \qquad (2.10)$$

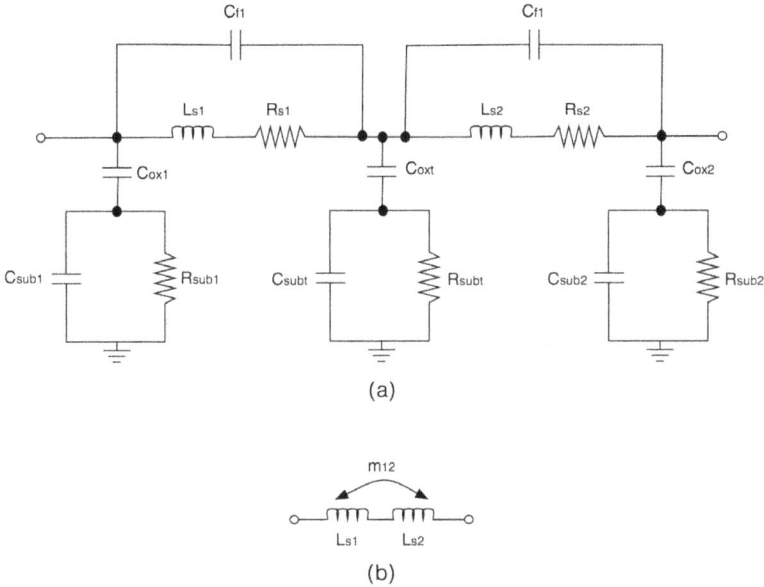

FIGURE 2.3 (a) Model of two on-chip inductors in series. (b) Simplified model of two inductors in series.

To evaluate the coupling between integrated spiral inductors, electromagnetic simulations (EM) have been carried out using a commercially available planar EM simulator (Momentum). In order to get the simulator to generate useful data, the substrate definition was first calibrated. This was accomplished by using measured data from previously fabricated inductors as a reference and then adjusting the substrate definition until it produced closely correlated data.[6,7,8]

Figure 2.4 shows the layouts of the simulated structures: L_{T1}, and L_{T2}. In the first case the inductors are coiled in the same direction, whereas in the latter case both inductors are coiled in opposite directions. The simulated quality factor and inductance of both structures along with the isolated inductor response (L_P) are depicted in Fig. 2.5. As observed, in the case where the spirals are coiled in the same direction, the total inductance is around 7 nH at 4 GHz, that is, more than two times the isolated inductance. Conversely, in the case where the spirals are coiled in opposite directions, the total inductance is

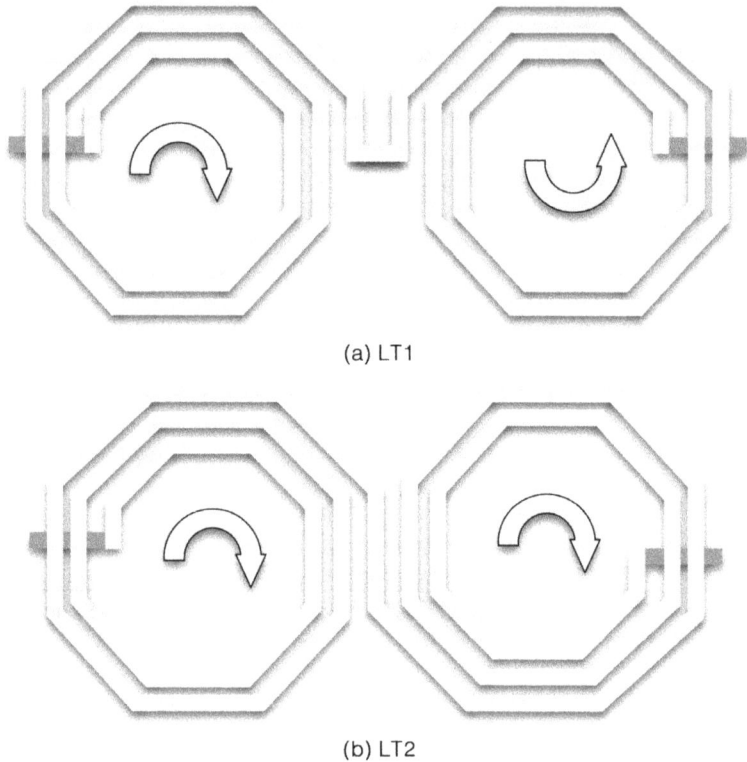

(a) LT1

(b) LT2

FIGURE 2.4   Layouts of series inductors coiled in the same way (a) and in the opposite way (b).

6.4 nH at the same frequency, that is, less than the sum of both inductances.

Regarding the quality factor, the situation is the same. L_{T1} shows a higher Q than L_{T2} because both structures share the same series resistance but their inductances are different.

From the results shown in Fig. 2.5, it can be stated that, in the design of DAs where inductors are situated close to each other to reduce the chip area, it is important to correctly orient the inductors to minimize the effect of mutual coupling.

2.3.2   Stacked Inductors

Area reduction can also be achieved by adopting a multi-level or stacked structure (MLS)[7,9,10,11,12,13] (see Fig. 2.6), instead of the

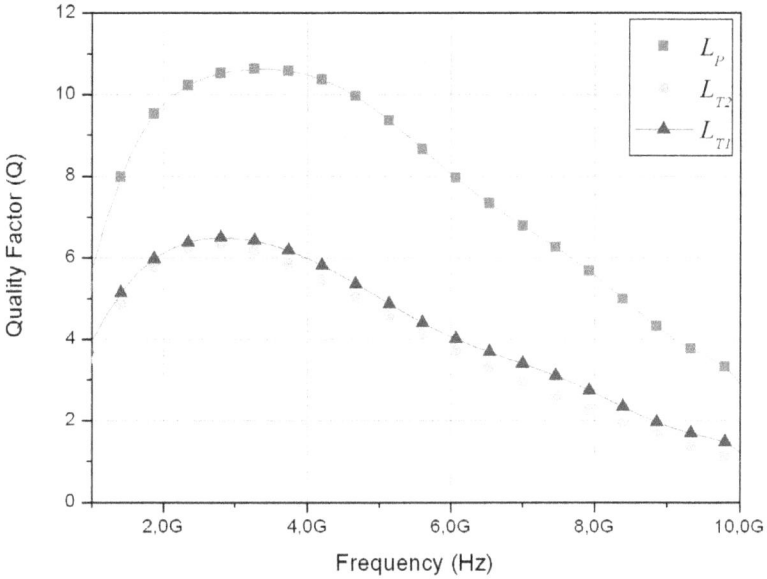

(a)

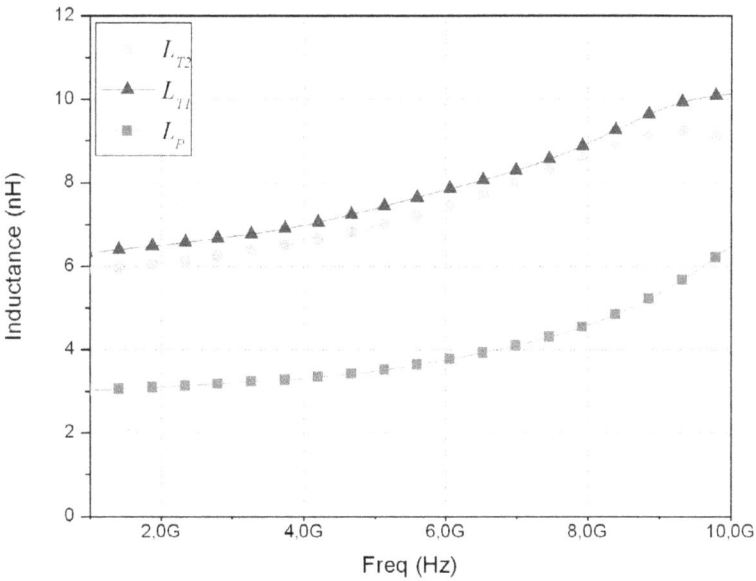

(b)

FIGURE 2.5 Influence of mutual coupling in series connected inductors over quality factor (a) and inductance (b).

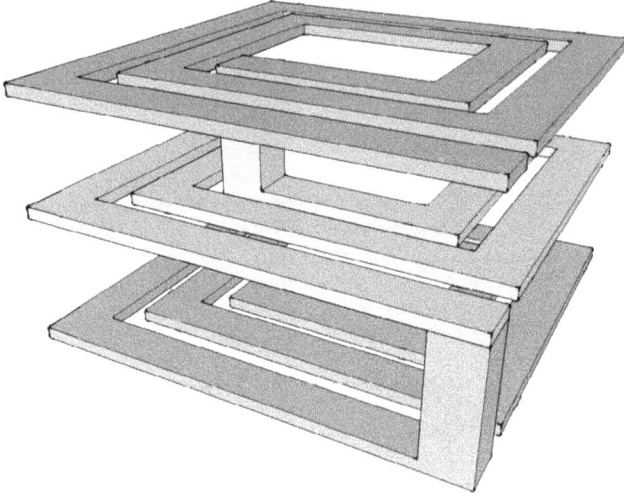

FIGURE 2.6   Stacked inductor layout.

usual single-level structure (SLS). The main difference between these two possibilities is, in the SLS case, the inductor is made with one metal layer, usually the top one because it is thicker than the lower metal layers, and in the MLS the turn is expanded vertically. Thus, for a fixed geometry, the multilevel structure presents a bigger length than the single-level one, and as a consequence, the inductance is increased. This is the main advantage of the multilevel structure: the same inductor values can be obtained occupying less area than with the single-level one.

The analysis of the type of inductors is the same for the MLS and SLS. This is because, from the physical point of view, the metal-to-metal and the metal-to-substrate capacitances are larger in the MLS than in the SLS. Therefore the lumped-element equivalent circuit that describes the behavior of the SLS can be applied to characterize the MLS.

In order to compare both structures, Fig. 2.7 shows the inductance and quality factor of two 1-nH SLS and MLS inductors. As the figure suggests, the quality factor and the self-resonant frequency f_{SR} of stacked inductors are lower than that of planar spiral inductors.[7]

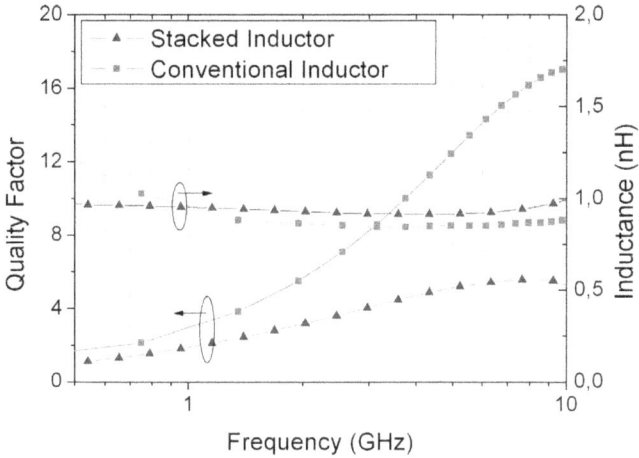

FIGURE 2.7 Conventional inductor vs. stacked inductor.

2.4 Experimental Results

In order to validate the area reduction techniques, three prototypes have been fabricated in a low-cost CMOS 0.35 μm process. The circuits are called DA1, DA2, and DA3, and they correspond, respectively, to the conventional design, compact design, and compact design with stacked inductors. The microphotographs of the fabricated circuits are shown in Fig. 2.8.

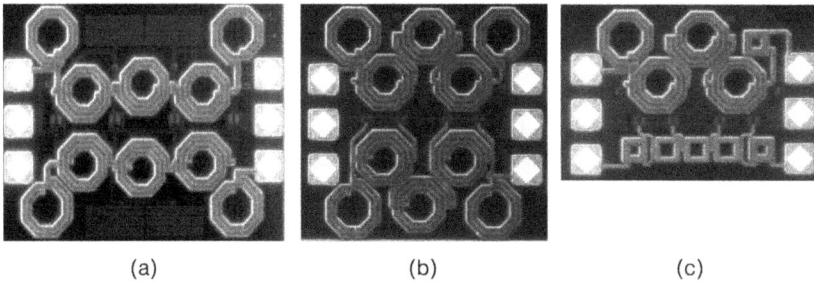

(a)   (b)   (c)

FIGURE 2.8 (a) DA1: conventional design (0.7 mm^2), (b) DA2: compact design (0.6 mm^2), (c) DA3: compact design with stacked inductors (0.4 mm^2).

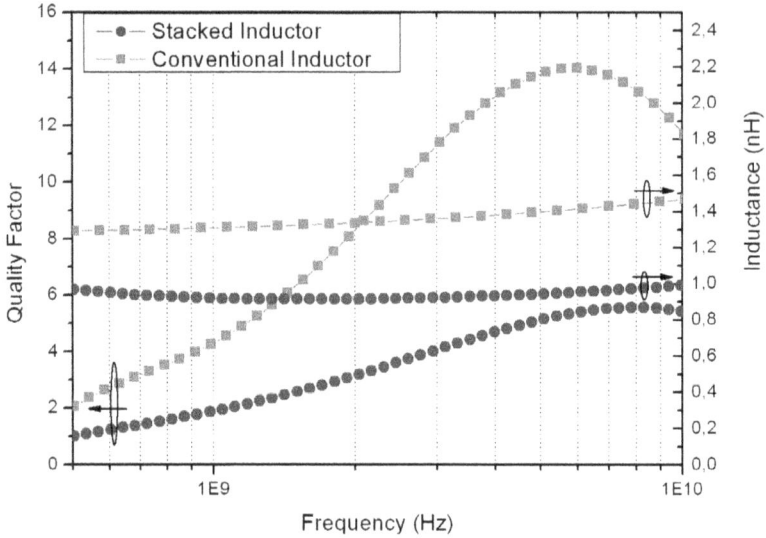

FIGURE 2.9 Inductors employed in the DA.

The design issues concerning the implementation of the three circuits are described in the following paragraphs.

One of the difficulties in designing a fully integrated DA is creating the required high-quality inductors. Figure 2.9 shows the simulated quality factor and inductance of the inductors used in the three designs. Table 2.2 shows the geometrical parameters of the chosen coils. The first one is a conventional planar spiral inductor and the second is a stacked inductor. The conventional inductor was used to generate $L_d = L_g$ and $L_d/2 = L_g/2$ in DA1 and DA2. As observed in Fig. 2.9, at 5 GHz this inductor presents a quality factor and an inductance of 9.28 and 1.3 nH, respectively. Although this inductor has less than half of the inductance of $L_d = L_g$, its physical layout perfectly matches the other components in both designs. On the other hand, the stacked inductor was used in the gate line of DA3.

	s(μm)	n	r(μm)	W(μm)
Conventional	2	2.5	100	16
Stacked	2	2×1.5	40	10

TABLE 2.2 Inductor's geometrical parameters.

This circuit keeps the conventional inductor in the drain line because the current flowing through this line is too high to be supported by stacked inductors.

Due to the very high frequency of operation, several GHz, special attention has been paid to the layout. Thus, enough design accuracy has been achieved by adding an accurate high-Q inductor model and optimizing parasitic effects coming from discontinuity and interconnection. The designed DAs utilize a transistor size of 130 μm (equivalent to 13 gates with 10 μm gate width) and a capacitance C_d of 150 fF.

The DA1 circuit occupies an area of 0.74 mm^2, which includes the pad frame. In contrast to the conventional design, the compact design DA2 occupies a total area of 0.61 mm^2, i.e. a 17% reduction. Finally, the compact design with stacked inductors DA3 occupies a total area of 0.47 mm^2, which implies a 36% reduction in area.

After the measurement of several samples, the frequency response is shown in Fig. 2.10. The power gain of DA1 is 6 dB with $\pm$0.6 dB flatness from 1 GHz to 5 GHz and a unity gain around 8.6 GHz. The input and output matching are better than -10 dB over the bandwidth. The increase in gain at low frequency is due to the higher impedance of the blocking capacitance at low frequency. All measurements were taken under identical DC bias conditions: 3 V on the drain and 0.8 V on the gate. At this bias point the DA consumes 30 mA, giving a total power dissipation of 90 mW. Finally, Figure 2.10 also shows the noise response. The noise figure is under 5 dB from 1 GHz to 6.5 GHz, and it is around 7.5 dB at 8.5 GHz.

The frequency response of DA2 is approximately the same of DA1. Regarding the noise figure, DA2 performance is better than DA1 mainly because the parasitics associated with the connection tracks have been reduced.

With respect to DA3, in spite of the stacked inductor in which performance is worse than the conventional ones, its response is very similar to DA1 and DA2. The noise figure is a little bit higher than that of the conventional design. This is because to the series resistance associated with stacked inductors is larger than that of conventional inductors.

Table 2.3 summarizes the performance of the presented amplifiers, with a comparison to previously published DAs.

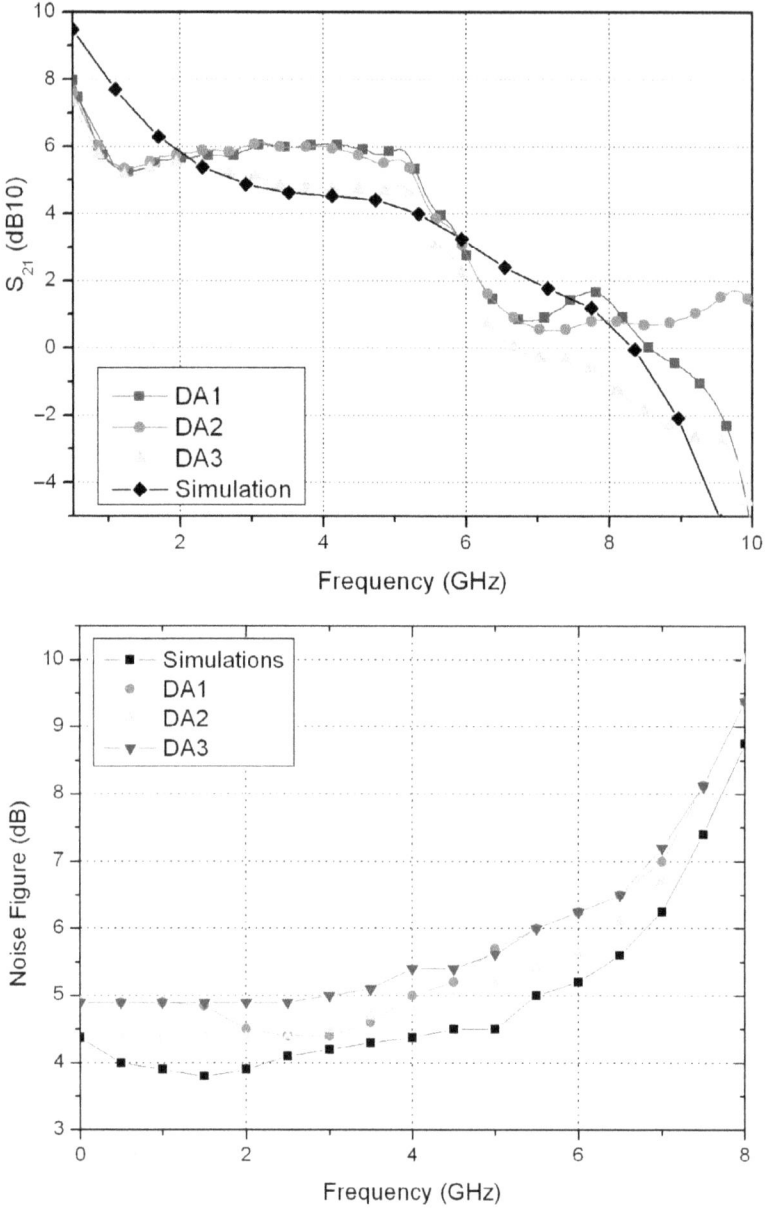

FIGURE 2.10 S_{21} and noise figure measurements.

Ref.	Gain (dB)	BW (GHz)	NF (dB)	P_{1dB} (dBm)	Area (mm²)	ft^* (Tech.)	P_{DC} (mW)	FOM¹	FOM²
4	6.1	5.5	6.8	8.8	1.12	10.5 (0.6μ)	83.4	151	132
14	5.5	8.5	10.85	N/A	2.86	10.5 (0.6μ)	286	57	–
15	7.3	22	5.2	10	1.6	33.7 (0.18μ)	52	108	95
15	10.6	14	4.35	5.3	1.35	33.7 (0.18μ)	52	124	106
16	6	27	6	10	1.62	33.7 (0.18μ)	68	107	94
17	4	8	5.4	8	0.84	33.7 (0.18μ)	23	208	182
18	10	11	4.6	N/A	1.44	33.7 (0.18μ)	19.6	119	–
18	16	11	4.5	N/A	1.44	33.7 (0.18μ)	100	110	–
DA1	7	6.5	5	12.3	0.74	8.13 (0.35μ)	90	231	207
DA2	7	6.5	4.5	12.4	0.61	8.13 (0.35μ)	90	282	253
DA3	5.5	6.5	6	11.2	0.47	8.13 (0.35μ)	90	364	325

[1]FOM not including P_{1dB}, [2]FOM including P_{1dB}.

TABLE 2.3 Summary of LNA performance and comparison with previously published designs.

To provide an objective method to compare the developed circuits and other similar works, a figure of merit (FOM) has been used:

$$FOM = \frac{P_{1dB}}{P_{noise}} \frac{1}{P_{DC}} \frac{f_h}{f_t^*} \frac{1}{AREA} \qquad (2.11)$$

This expression includes the DC power consumption (P_{DC}) and output noise power ($P_{noise} = P_{th}FGain$), where $P_{th} = kT$ is the thermal noise floor given by -174 dBm/Hz at T = 290 K. In addition, in order to quantify how efficiently the available bandwidth of the technology is utilized, a relative measure for bandwidth is introduced through the f_h/f_t^* factor, where f_h is the upper LNA corner frequency and f^* is the technology unity current gain bandwidth (f_t) around the maximum of the product (g_m/I_D) f_t. Finally, AREA is the area occupied by the circuit, and it allows comparison of the designs in terms of area consumption. The proposed FOM includes the output 1-dB compression power (P_{1dB}) as a measure for linearity. However, some authors do not include this measurement, and as a consequence, two FOMs have been plotted in Table 2.3: one including the P_{1dB} and the other one without any linearity reference.

The DA presented in Ballweber et al.[4] has an excellent FOM, mainly because it utilizes very efficiently the available bandwidth of the technology. However, the area of this circuit almost doubles our designs, and as a consequence, its FOM is lower than ours. The same authors utilize a fully differential topology in[14] to achieve a wider bandwidth than its single-ended counterpart. However both, area and power consumption double and the achieved FOM is low.

The works of Liu et al.[15] exhibit both higher gain and bandwidth than our designs. Also the power dissipated is low being the NF similar than our designs. However, our DAs achieve better FOMs mainly because they utilize more efficiently the available area.

The low power techniques presented in[17,18] use a low P_{DC} to achieve low noise figure and good gain, but they are, however, fundamentally limited by a large area.

Finally the DA reported in Amaya et al.[16] uses coplanar waveguides to implement the required inductances. This technique achieves a very high frequency of operation but at the cost of a very large area.

Regarding the presented designs, the best FOM is achieved, as expected, by DA3. This design employs stacked inductors to reduce area. This kind of inductor occupies less chip area than that of planar spiral inductors since the turn is expanded vertically. Usually, the top metal is thicker than lower metal layers, and thus the Q of stacked inductors is lower than that of planar spiral inductors. However, because the area occupied is much smaller, substrate losses are smaller, so only little performance degradation of the stacked inductor circuit is achieved over the planar spiral inductor one. The performance degradation is only a little less in the stacked inductor circuit than in the planar spiral one, so it is possible to reduce the area with a minimum influence on the circuit response.

2.5 Conclusions

This chapter presented a first approach to LNAs for UWB communications. Distributed amplifiers are the most classical way to implement amplifiers with a huge bandwidth and a relative gain.

The main drawbacks of this structure are the high power consumption and the elevated area. In order to reduce the area, two techniques were reported in this chapter. The first one reallocates the drain and gate line inductors while minimizing the mutual inductance between them. The other technique employs stacked inductors. Although the quality factor of the stacked inductor is lower than the planar inductor, because the area is much smaller, substrate losses are also smaller, and only little circuit performance degradation is achieved when stacked inductors are used. Using the above techniques, three fully

Design	Gain (dB)	BW (GHz)	NF (dB)	P_{1dB} (dBm)	Area (mm^2)	P_{DC} (mW)
DA1	7	6.5	5	12.3	0.74	90
DA2	7	6.5	4.5	12.4	0.61	90
DA3	5.5	6.5	6	11.2	0.47	90

TABLE **2.4** Distributed amplifier's specifications.

integrated distributed amplifiers have been designed, fabricated, and tested. Table 2.4 shows a summary of their specifications.

The next chapter explores other alternatives to implement low-noise amplifiers for ultra-wideband communications that try to reduce the power consumption and the occupied area.

CHAPTER 3

Wideband Low-Noise Amplifiers

3.1 Introduction

Distributed amplifiers, presented in the previous chapter, exhibit a high power consumption and occupy a considerable area. This chapter describes alternatives to low-noise amplifiers that reduce the power consumption and area.

3.2 Wideband Low-Noise Amplifier

3.2.1 Narrowband Inductively Degenerated Amplifier

In this section, the typical narrow band inductively degenerated amplifier configuration is presented because it is the base of the wideband LNA.

Figure 3.1 shows the typical schematic of a narrowband LNA. The input transistor (Q_{CAS1}) is in common emitter configuration, and it is the main contributor to the circuit noise. The noise figure of the LNA depends directly on the emitter area and on the bias of Q_{CAS1}. The cascode stage, composed of Q_{CAS1} and Q_{CAS2}, reduces the Miller capacitance, decreasing the effective base collector capacitance (C_{bc}) of Q_{CAS2}. This makes the amplifier unilateral, that is with low S_{21}.

A requirement of many communication systems is to prevent leakage of local oscillator power from the mixer back to the antenna.[19] The cascode also enhances the overall gain by increasing the output impedance. The resonant circuit composed of L and C is the load of the cascode stage. This allows a high gain with a low voltage supply. The tank resonant frequency is adjusted to the frequency of interest (ω_0).

The noise in a transistor is proportional to the transistor small signal transconductance $g_m = 1/re = I_C/V_T$ (where V_T is

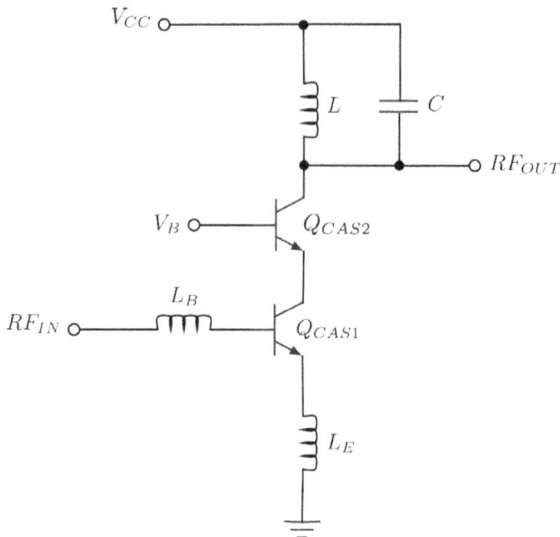

FIGURE 3.1 Simplified schematic of the LNA with inductive degeneration.

the thermal voltage and I_C is the collector current) and the transistor base and emitter resistances, r_b and r_e. To minimize r_b, the transistor must have a large area, and to maximize g_m, I_C must be high. If the transistor area is increased, the input capacitance will also increase. This will attenuate the input signal and will raise the NF. As a result the NF will reach a minimum for a particular combination of area and polarization current.

The next step in minimizing the noise is matching the LNA input. The antenna output impedance is 50 Ω, and through inductive degeneration it is possible to match the input and achieve an excellent tradeoff between conjugate matching and minimum noise. The inductive degeneration consists of introducing a series inductance (L_E) at the emitter as shown in Fig. 3.1. The inductance value is approximately given by Eq. 3.1:

$$L_E \approx \frac{Z_0}{\omega_T} \tag{3.1}$$

The higher the transistor $\omega_T = g_m/C_i$, the lower the value of L_E needed for matching, and the lower the amount of noise added to the LNA by the series resistance of the inductor. L_E changes the real part of the input impedance, and to modify the imaginary part another inductor L_B is introduced as shown in Fig. 3.1. An expression of the noise factor for the LNA with inductive degeneration that takes into account the above discussion is shown in Eq. 3.2:[5]

$$F = 1 + \frac{R_b + R_e}{Z_0} + \frac{g_m}{2} \cdot Z_0 \cdot \left(\frac{\omega_0}{\omega_T}\right)^2 \tag{3.2}$$

Alternatively this equation can be expressed as:

$$F = 1 + \frac{r_b + r_e}{Z_0} + \frac{1}{2 \cdot g_m \cdot Z_0 \cdot Q^2} \tag{3.3}$$

where Q is the quality factor of the input matching network. The noise factor improves with a higher Q because more voltage gain is seen across the input capacitance of the transistor. The input impedance is resistive only in a narrow bandwidth (ω_0/Q) around the resonance frequency ω_0. To obtain a wideband impedance matching, the Q of the matching circuit should be significantly lowered. This will largely degrade the noise figure, which defeats the purpose. As a result, this type of amplifier cannot be used for wideband applications.

3.2.2 Wideband Inductively Degenerated Amplifier

Wideband impedance matching expands the use of an inductively degenerated amplifier, by embedding the input network of the amplifying device in a multisection reactive network so that the overall input reactance is resonated over a broad bandwidth. In this way, a wideband input match is achieved, and at the same time, good noise performance is attained. In the proposed design, shown in Fig. 3.2, a fourth-order doubly terminated band-pass filter is used to resonate the reactive part of the input impedance over the whole band. As long as the upper and lower cutoff frequencies (ω_U and ω_L) of the filter are far from each other, this fourth-order band-pass filter can be seen as a combination of two filter sections, one in a low-pass configuration and the other one in a high-pass configuration.

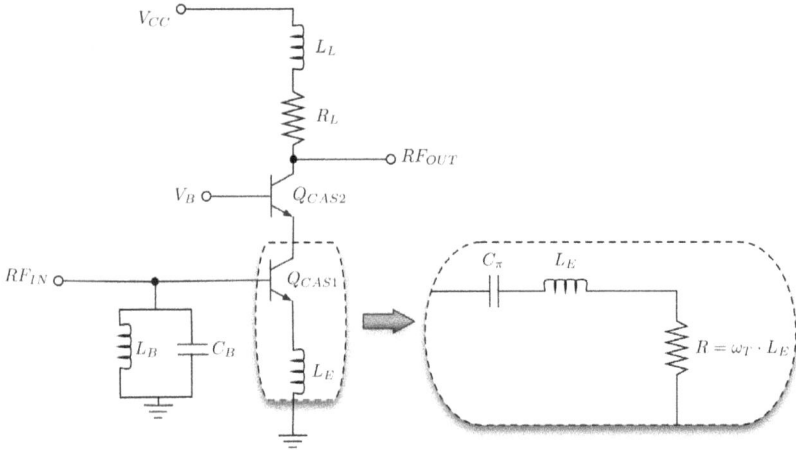

FIGURE 3.2 Simplified schematic of the LNA with wideband impedance matching and wideband load.

The high-pass filter section is composed of L_B and C_π and its cutoff frequency is given by:

$$high\text{-}pass\left\{L_B = \frac{R}{\omega_L}; C_\pi = \frac{1}{\omega_L R}\right. \tag{3.4}$$

On the other hand, the low-pass filter section is composed of L_E and C_B and its cutoff frequency is given by:

$$low\text{-}pass\left\{L_E = \frac{R}{\omega_U}; C_B = \frac{1}{\omega_U R}\right. \tag{3.5}$$

These two circuits provide an input impedance equal to R in the pass-band between ω_U and ω_L.

To provide a wideband operation, one might suggest replacing the resonant load in the narrowband circuit by a resistor. However, this would lead to gain response falling with the frequency because of both the pole generated by the resistor load (R_L) and the capacitance of the output node (C_{OUT}). A technique commonly used to increase the bandwidth is to replace the load resistor by a shunt peaking resistor[5] composed of L_L and R_L (see Fig. 3.2). The addition of an inductance in series with the load resistor provides an impedance component that increases with frequency (i.e., introduces a zero), which helps decrease the impedance of the load capacitance, leaving a net impedance that remains roughly constant over a broader frequency range than that of the original RC network. R_L should be sufficiently low so that the inductive region of the impedance spans the pass-band.

With this configuration the inductive load equalizes the voltage gain to a constant value across the pass-band. The problem is that C_{OUT} introduces a spurious resonance with L_L (peaking), which must be kept out-of-band. As long as C_{OUT} represents all the loading on the output node, including the transistor output capacitance, the loading by interconnect, and subsequent stages and parasitic capacitances of the inductor, all these contributions should be minimized to ensure self-resonance beyond ω_U.

3.2.3   Wideband Low-Noise Amplifier Design

Figure 3.3 shows the designed wideband LNA circuit. For measurement purposes, an emitter follower buffer is included to drive an external 50 Ω load. C_2 and C_5 capacitors are for AC decoupling, and R_{BIAS} has a large value in order to bias the output buffer with V_{BIAS1} voltage.

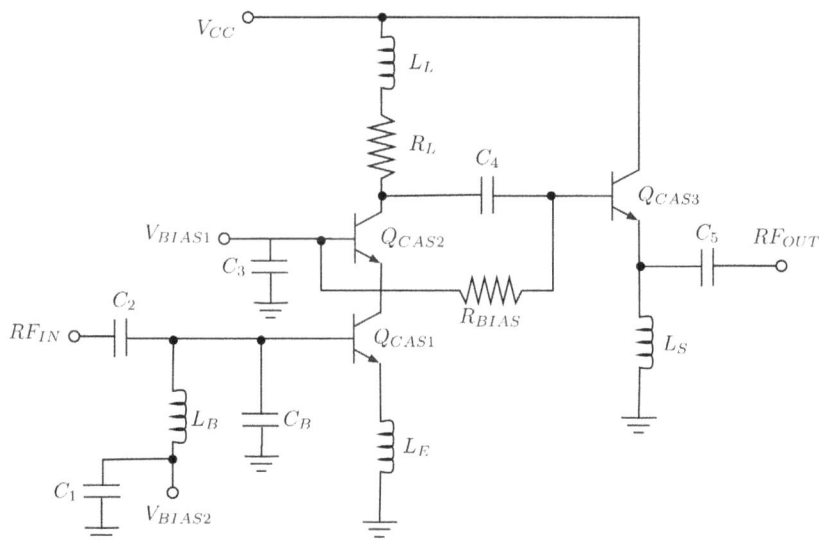

FIGURE 3.3   Schematic of the wideband LNA.

The performance of a narrowband LNA is determined by the limited quality factor of the integrated inductors.[20] Its optimization relies on achieving the highest Q for a given inductance value at the frequency of operation. In the case of a wideband operation, this assumption is not convenient and a further analysis is needed. In shunt peaking applications, the biggest issue is the reduction of bandwidth because of additional parasitic capacitance introduced by on-chip inductors. As a consequence, prior to their use in any circuit, spiral planar inductors must be modelled accurately over a wide range of frequencies. In this work, the analytical model proposed in Goni et al.[8] has been used to implement an optimization algorithm that provides the geometry of the inductor with the best quality factor for a given inductance value and frequency of operation. For every inductor of the circuit, a set of spirals with the same inductance but optimized for different frequencies were simulated. In order to better predict the inductor performance, S parameters from full-wave electromagnetic simulations were used to simulate the circuit. A commercially available planar EM simulator, Momentum, was used to predict the broadband response of inductors. The inductors were designed using the top-level metal, which is thicker and more conductive than the rest. All of these inductors were laid out in an octagonal fashion, with external radio (r) up to 170 μm, metal width (w) between 5 and 25 μm, and up to 5.5 turns (n). The spacing between tracks is fixed to the minimum allowed by the technology, 2 μm, in order to maximize the inductance value.

The simulated quality factor and inductance of the spirals are reported in Figs. 3.4 to 3.7. In these figures, the wideband LNA simulated power gain using the above spirals is also plotted.

As can be seen the circuit performance is insensitive to inductors L_B and L_E quality factors. With maximum quality factor frequencies (f_{Qmax}) for L_B ranging from 2 to 5 GHz, the gain remains unaltered. Only little variations are observed at low frequency and must be attributed to the slight differences in inductance value. In the case of L_E, the inductance value is low, and the geometric characteristics of the simulated spirals are similar. As a result, the gain flatness is not affected.

On the other hand, L_S is used for biasing purposes and does not affect the frequency response of the circuit. However, this

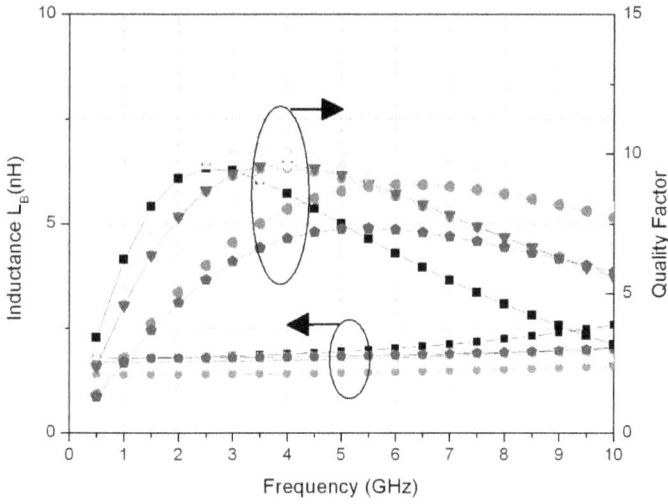

(a)

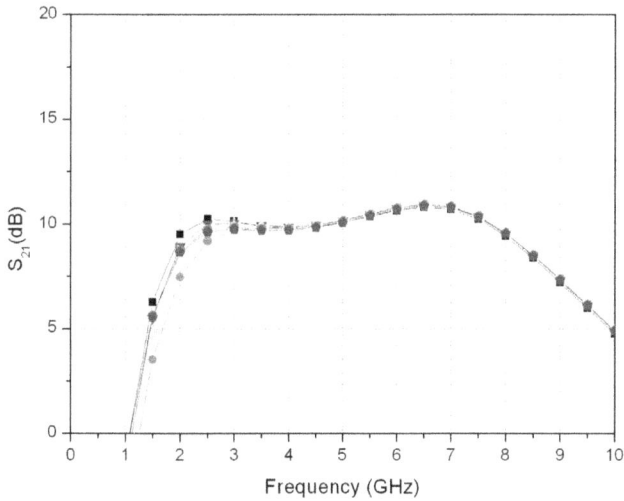

(b)

FIGURE 3.4 (a) Quality factor and inductance of spirals suitable for L_B. (b) Power gain for various L_B.

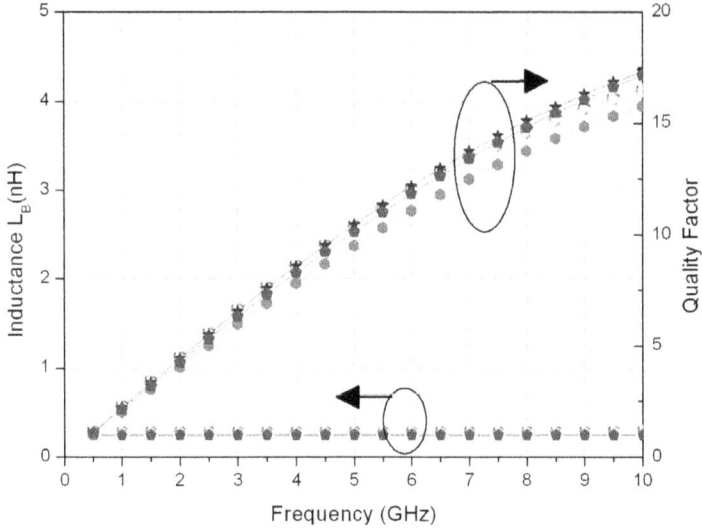

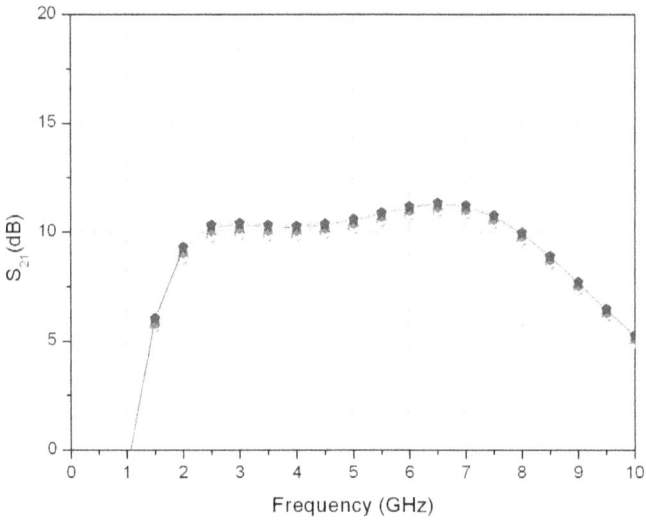

FIGURE 3.5 (a) Quality factor and inductance of spirals suitable for L_E. (b) Power gain for various L_E.

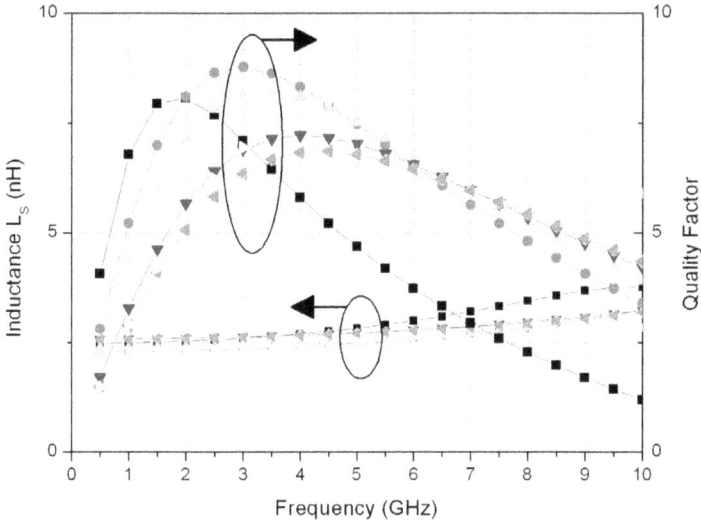

(a)

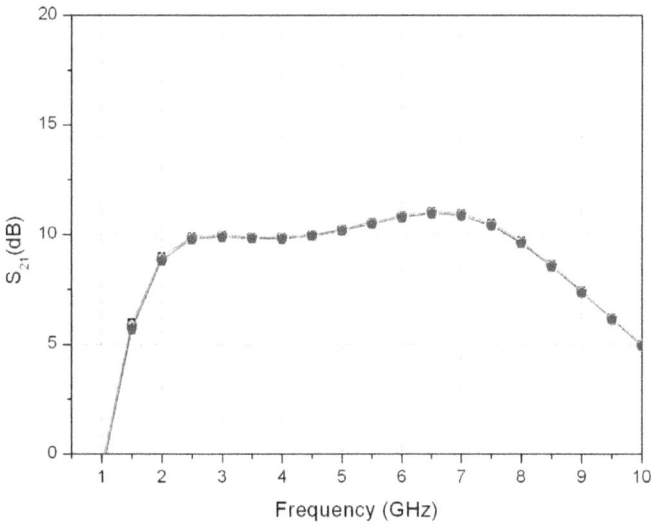

(b)

FIGURE 3.6 (a) Quality factor and inductance of spirals suitable for L_S. (b) Power gain for various L_S.

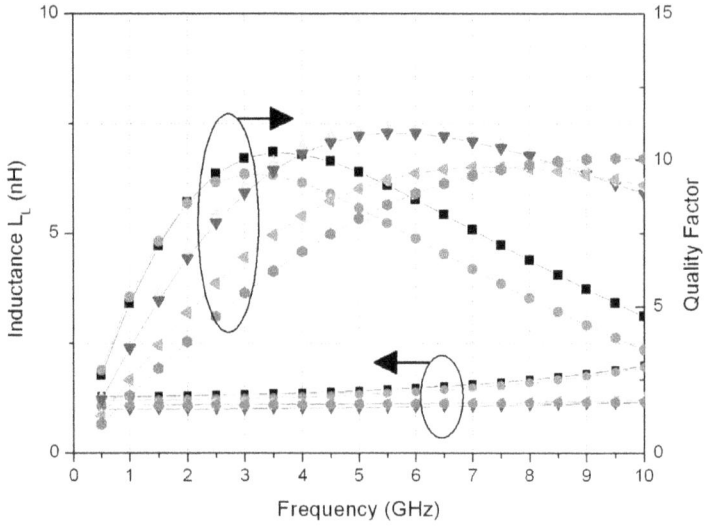

(a)

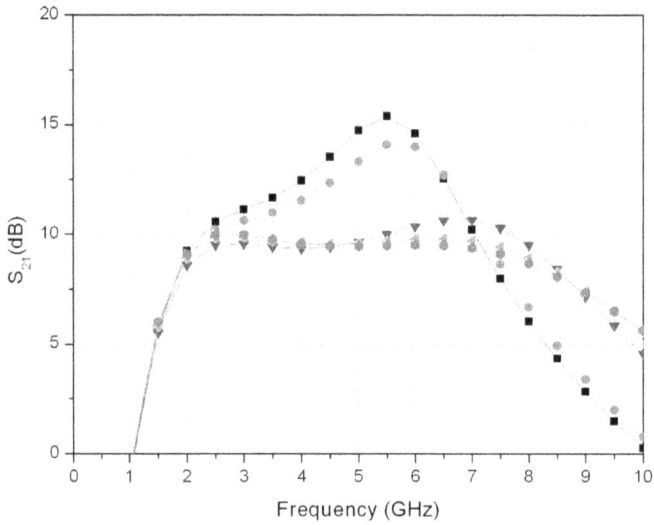

(b)

FIGURE 3.7   (a) Quality factor and inductance of spirals suitable for L_L. (b) Power gain for various L_L.

result no longer holds for the case of L_L. As can be seen in Fig. 3.7, the gain flatness is strongly affected by the inductor Q. The best results are obtained for inductors that exhibit an equalized quality factor through the entire band, despite having Q with an irregular shape through the band of interest.

3.2.4 Experimental Results

Figure 3.8 shows the final wideband LNA photography. The total chip size, including probe pads, is $665 \times 665 \ \mu m^2$. The amplifier draws 5.3 mA and the output buffer draws 6.5 mA from a 3.3 V supply.

The layout has been implemented using a 0.35 μm BiCMOS process. The amplifier was measured on wafer using a Cascade SUMMIT 9000 probe station, 35 GHz probes, and 20 GHz Agilent 8720ES vector network analyzer. The probe pads were octagonal, optimized for RF. Three ground-signal-ground (GSG) and one signal-ground-signal (SGS) pad structures with 150 μm pitch were used, as depicted in Fig. 3.8.

Figure 3.9 shows measured and post-layout simulated S parameters of the wideband amplifier, which agree quite well, except for the S_{22}, which is better in measurements. This is due to parasitic resonance of the output buffer, which was not taken

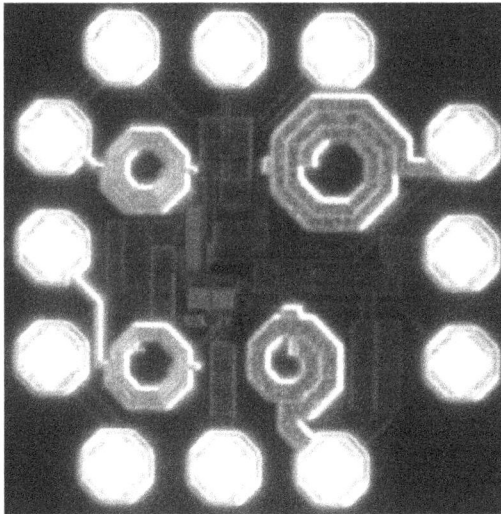

FIGURE 3.8 Wideband amplifier layout.

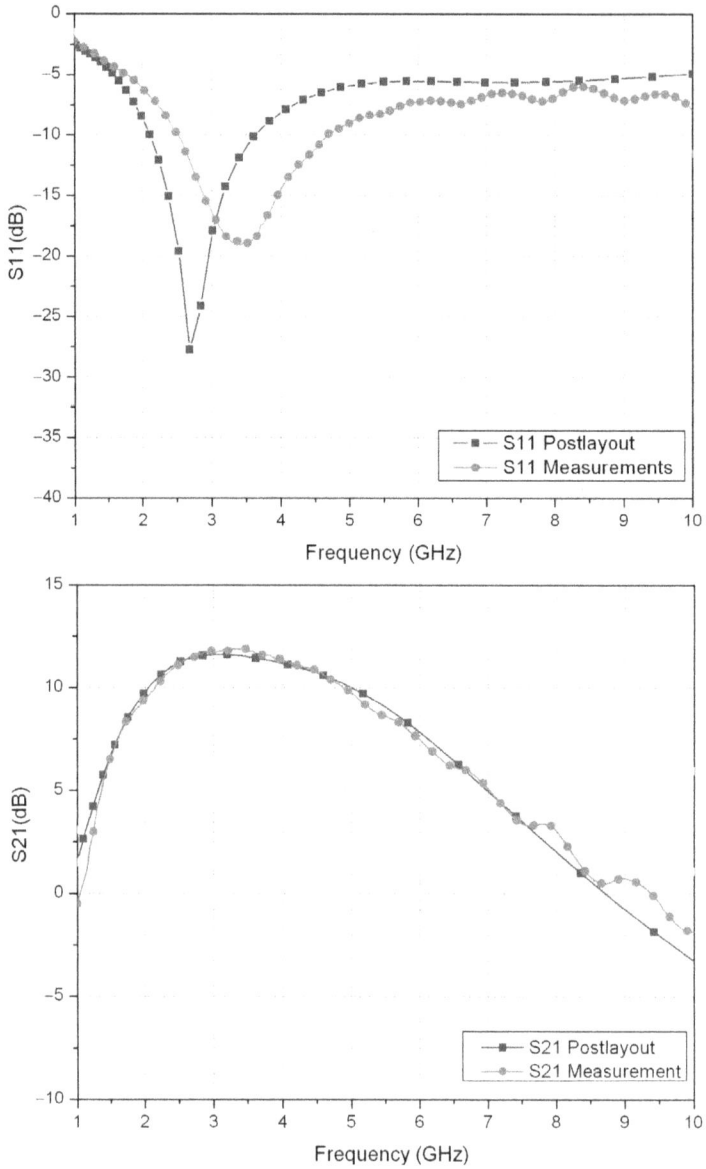

Figure 3.9 Measured and simulated scattering parameters for the designed wideband LNA. (*continued*)

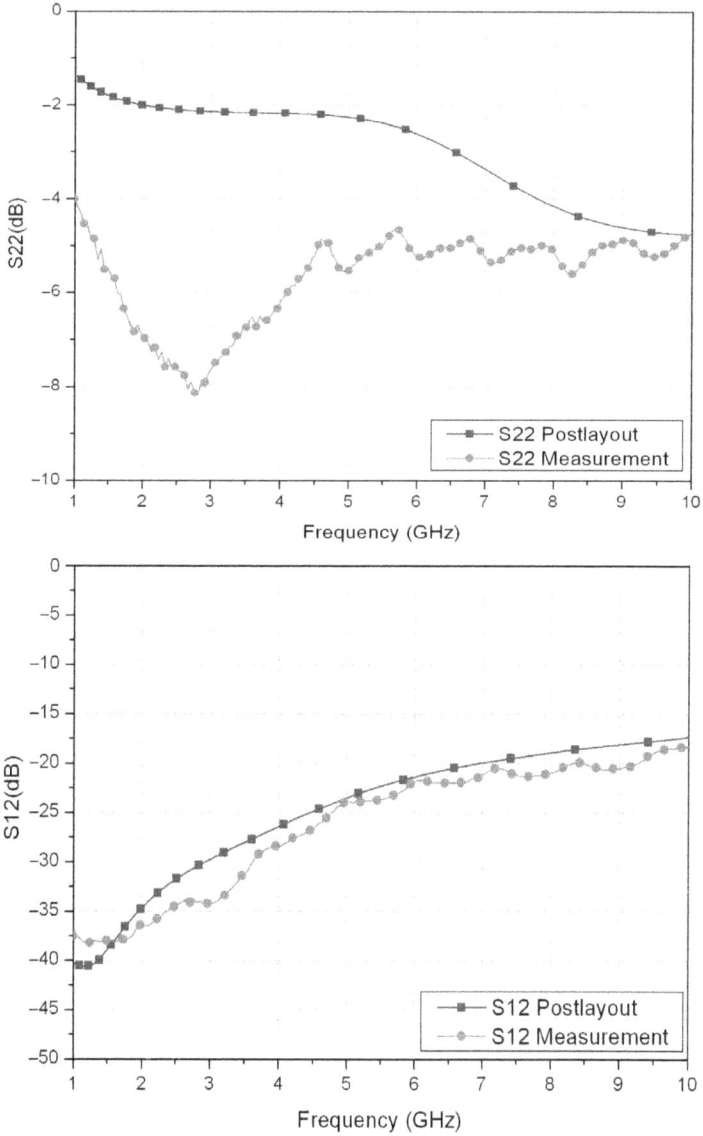

Figure 3.9   (*Continued*)

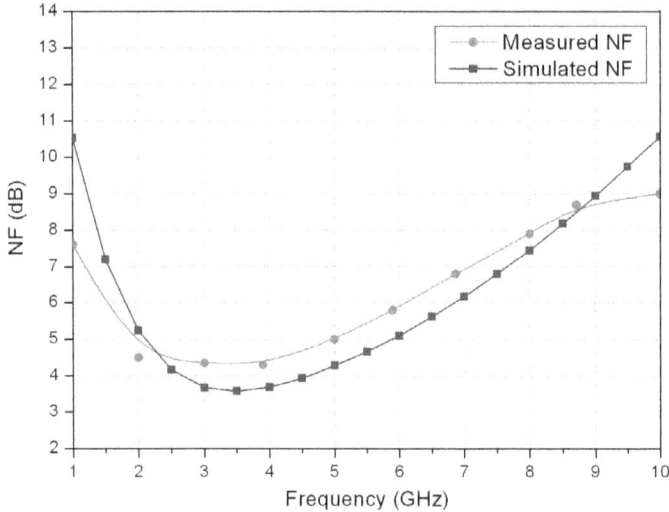

Figure 3.10 Measured and simulated NF for the designed wideband LNA.

into account in the simulations. As expected, a maximum power gain of 12.5 dB was achieved at 3.4 GHz with a −3 dB bandwidth of 1.7–5.3 GHz. A unity gain was measured at 9.4 GHz. The measured input return loss (S_{11}) is lower than −5 dB over the bandwidth. The output return loss (S_{22}) has a maximum value of −4 dB due to the source follower output stage. The reverse isolation (S_{12}) is greater than 23 dB due to the cascode stage. The NF measurement was done in a noise-free environment with an E4440 Agilent 26.5 GHz spectrum analyzer and a 346 C noise source. Figure 3.10 shows the amplifier measured and simulated NF. The NF varies from 4.3 dB at 3.9 GHz, to 5.2 dB at 5.3 GHz. This result shows good agreement between measured and simulated data. The IIP3 was measured as −4 dBm.

Measured results and a brief comparison with similar amplifiers are summarized in Table 3.1, which shows that the designed LNA, using an cheap technology, has a good tradeoff between bandwidth, noise figure, gain, linearity, and power consumption. Keep in mind that although the power consumption of our amplifier is the highest, the circuit was fabricated in an older technology compared to the other publications.

Ref.	BW 3 dB (GHz)	Max. gain (dB)	Max. NF (dB)	IIP3 (dBm)	P_{DC} (mW)	Techn.	Year
21	3.1–10.6	9.18	7.2	7.25	23.5	0.18 μm	2007
22	2.0–4.6	9.8	5.2	−7	12.6	0.18 μm	2005
23	3.1–4.8	15	4.9	−2.2	20	0.25 μm	2006
24	3.0–5.0	12.7	5.02	−9.7	16.4	0.18 μm	2005
25	3.1–7.5	19.1	3.8	−2.2	32	0.18 μm	2006
26	3.0–5.0	12	4.5	−	20	0.18 μm	2009
This work	1.7–5.3	12.5	5.0	−4	32	0.35 μm	2011

TABLE 3.1   Comparison with published wideband amplifiers.

3.3 Flatness Improvement

One of the main drawbacks of the wideband amplifier is the gain flatness. Usually, to extend the gain bandwidth, the load is composed of a shunt peaking resistor. This technique imposes an upper limit to maximum gain and flatness. To overcome this issue, a modification of the conventional shunt peaking is presented. In this case a CMOS technology has been used.

3.3.1 Circuit Description

The schematic of the wideband input matched CMOS LNA is shown in Fig. 3.11. As in the previous circuit, it consists of a wideband input matching circuit, a gain stage with inductive degeneration (L_g and L_s), and a wideband output load. In order to buffer the output to an external 50 Ω load, an emitter follower (M3) has been included. As the figure shows, the input matching circuit consists of a filter embedded with the input impedance of M1. In this case, a third-order band-pass Chebyshev filter in T configuration was selected. In order to increase the flexibility of the filter, C_P is introduced between the gate and source of M1.

The gain stage is composed of a cascode stage where the width and polarization current of the transistors are optimized for noise and power consumption.

Figure 3.12 shows the typical shunt peaking resistor load used to provide a wideband operation.[5] With this configuration the overall amplifier gain should be flat across the pass-band.

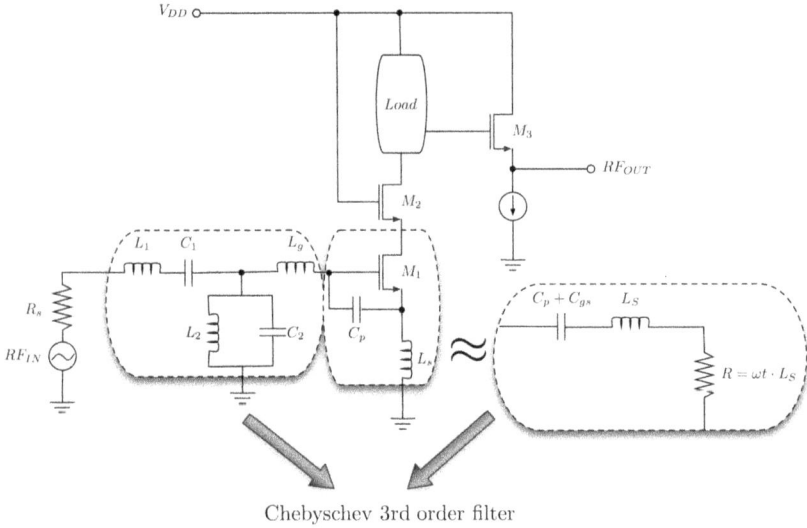

FIGURE 3.11 Wideband LNA simplified schematic with wideband input impedance matching.

The amplifier gain is given by the product of the transistor transconductance (g_m) and the magnitude of the impedance of the shunt peaking load, given by:

$$Z_L(j\omega) = \frac{R_L + \omega L_L}{1 - \omega^2 L_L C_{out} + j\omega C_{out} R_L} \tag{3.6}$$

where C_{out} represents the equivalent capacitance at the output node, which includes the transistor output capacitance,

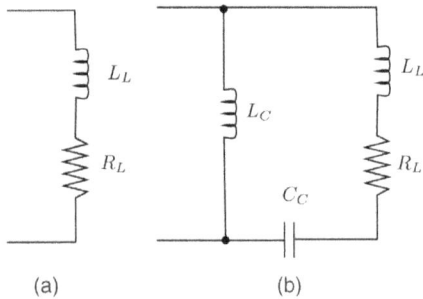

FIGURE 3.12 Conventional shunt peaking load (a) and modified shunt peaking load (b).

the loading by interconnections and subsequent stages, and the parasitic capacitance of the inductor. This expression contains a zero and two complex poles. The extended bandwidth comes from the $|Z(j\omega)|$ increase due to the poles below the $L_L\,C_{out}$ resonance ($\omega_0 = 1/L_L \cdot C_{out}$) and to the zero ($\omega z = R_L/L_L$). Unfortunately, this leads to a peak in the frequency response, thus degrading the flatness. As explained above, a possible solution is to keep both resonances out-of-band by using a low value of L_L, which in turn implies a low gain.

To have a large gain, R_L should be chosen sufficiently high to improve the gain at lower frequencies. However, the voltage headroom imposes an upper limit to R_L and, as a consequence, to maximum gain and flatness. To overcome this issue, a modification of the conventional shunt peaking load is proposed.

The proposed shunt peaking load is shown in the Fig. 3.12 (b). It is based on a conventional shunt peaking resistor, decoupled from the cascode stage through a capacitor C_C. To bias the active stage, an inductance L_C is placed between V_{DD} and the M_2 drain. The impedance of the new shunt peaking load is given by:

$$Z(j\omega) = \frac{j\omega L_C R_L \left(\dfrac{j\omega L_L}{R_L} + 1\right)}{1 - j\omega^3 L_C L_L C_{out} - \omega^2 L_C R_L C_{out} + j\omega(L_C + L_L)}$$

$$(3.7)$$

The C_C value has been chosen high; consequently, its effect is neglected and it does not appear in Eq. 3.7. With this configuration, R_L can be chosen higher than in a conventional shunt peaking load, overcoming the voltage headroom limitation. The immediate consequence is that a flatness improvement is achieved.

3.3.2 Experimental Results

To demonstrate the practical viability of the proposed structure in CMOS technology, both the proposed topology and the conventional one have been applied to a 3.1 to 4.8 GHz wideband amplifier, based on a 0.35 μm standard CMOS process. Both circuits were optimized with the pads, and on-chip spiral inductors analyzed with the Momentum electromagnetic simulator.[27]

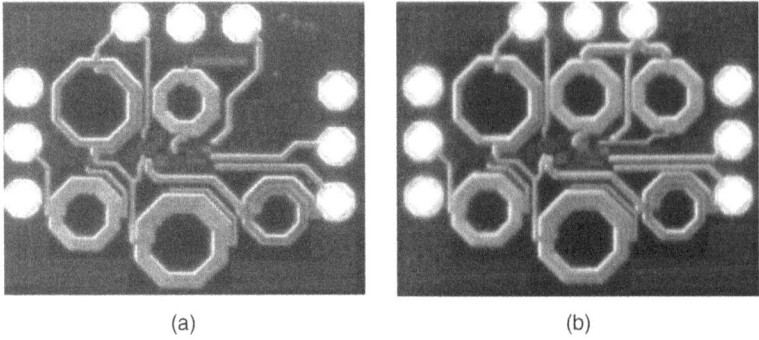

(a)   (b)

FIGURE 3.13 Photograph of the LNA with the shunt peaking (a) load and modified shunt peaking load (b).

Figure 3.13 shows the photos of two LNAs, one with shunt peaking load and the other with the modified shunt peaking load. As can be observed, the layouts are very similar with the exception of the L_C inductor in the upper right corner. In both designs the chip size, including the probe pads, is 949 × 760 μm. Each amplifier draws 17 mA from a 3.3 V supply.

The measured forward gain and input return loss of the amplifiers are shown in Fig. 3.14 for frequencies from 2 to

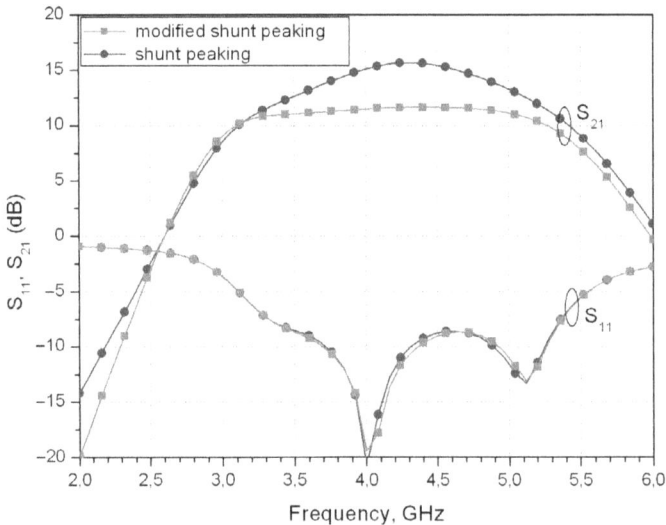

FIGURE 3.14 Measured S-parameters for LNA with shunt peaking load and modified shunt peaking load.

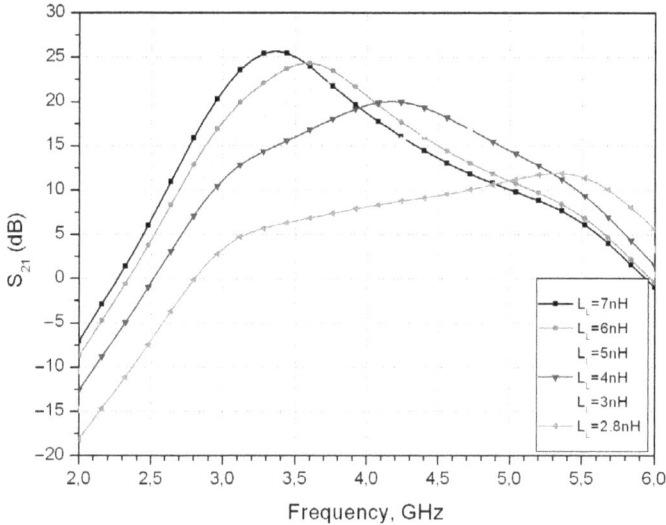

FIGURE 3.15 Gain simulation for different L_L inductance using conventional shunt-peaking.

6 GHz. For the proposed shunt peaking LNA, the power gain is fairly flat at approximately 10 dB for frequencies ranging from 3.1 to 5 GHz. However, the same cannot be said for the conventional case, where a peak is evident. To flatten the insertion gain, the zero pole frequency should be placed as close as possible to the upper edge of the band by lowering L_L. However, this entails a gain reduction, as can be seen in Fig. 3.15, where the simulated S_{21} is plotted for different ideal L_L inductors. As the inductance decreases, the flatness and bandwidth increase but the gain drops.

In both amplifiers, the input return losses remain the same, because both circuits share identical input matching circuits. As a consequence, the noise figure (see Fig. 3.16) is also the same in both cases.

3.4 Wideband Folded Cascode Amplifier

In previous sections, two different alternatives to implement wideband amplifiers based on cascode topology were presented. One of the drawbacks of the cascode amplifier is that this topology suffers from reduced linearity due to the stacking of two transistors, which reduces the available output swing.

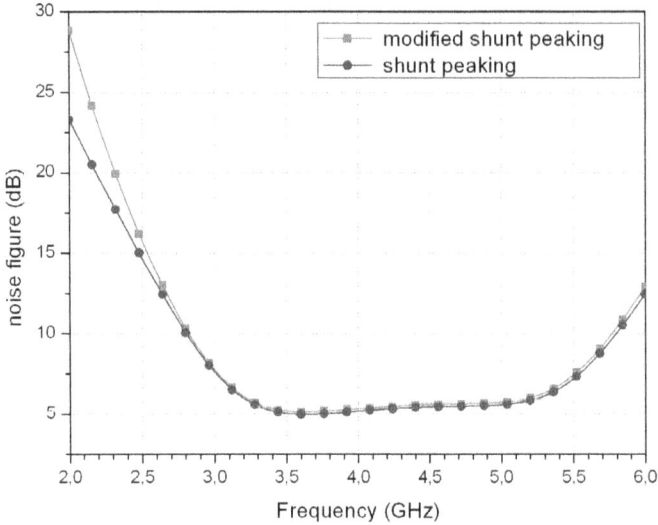

FIGURE 3.16 Measured noise figure for both LNAs.

To solve this problem, single transistor topologies are preferred for low-voltage operation such as the folded cascode LNA presented in this section.

3.4.1 Narrowband Folded Cascode Amplifier

Figure 3.17 shows the typical schematic of a narrowband folded cascode LNA. The input transistor (Q_1) is in common emitter configuration, and it is the main contributor to the circuit noise. The folded cascode stage is formed by Q_1 and Q_2, the resonant circuit is formed by L_L, and the output capacitance (C_{out}) is the load of the circuit. This folded structure permits a high gain with a low voltage supply.

To ensure that the circuit operates as a cascode amplifier, two conditions must be met simultaneously. First, to reduce the Miller effect, the signal gain at the collector of Q_1, relative to the input RF signal, should be near unity, and second, the entire RF signal current ($g_m \cdot |v_{be}|$) generated by Q1 should be fed into the emitter of Q_2. This is done by setting the L_C tanks (L_{T1} and L_{T2}) to resonate (i.e., have high impedances) at the frequency of interest. However, due to the finite quality factors of the integrated inductors, the impedances of the LC tanks

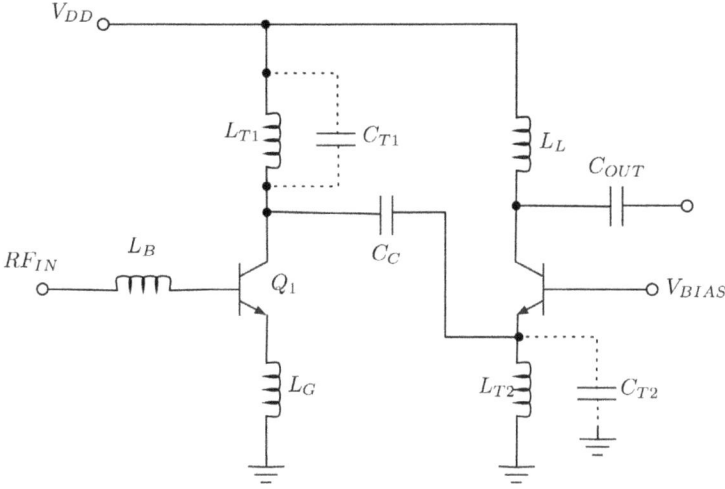

Figure 3.17 Simplified schematic of the LNA with inductive degeneration.

at resonance are also finite. They are given by the following equations:

$$R_{tank1} = (Q_{tank1} + 1) \cdot R_{L1} \tag{3.8}$$

$$R_{tank2} = (Q_{tank2} + 1) \cdot R_{L2} \tag{3.9}$$

where R_{tankn} and Q_{tankn} are impedances and the quality factor of the LC tanks at resonance, and R_{Ln} is the series resistance of the inductors. To avoid signal losses along the signal path, two considerations must be taken into account. To minimize signal divider losses, the LC tank impedance R_{tank1} must be much larger than the impedance looking into the coupling capacitor C_C. Similarly, to avoid signal losses to ground, the L_C tank impedance R_{tank2} must be larger than the impedance looking into the emitter of Q_2 (re_2). The above constraints are summarized as follow:

$$R_{tank1} \gg \frac{1}{j\omega C_C} + R_{tank2} || r_{e2} \tag{3.10}$$

$$R_{tank2} \gg r_{e2} = \frac{1}{g_{m2}} \tag{3.11}$$

The main benefit of using a folded cascode topology is its ability to operate at low supply voltages or, in other words, to exhibit a high linearity operation.

Linearity is an important parameter that specifies the ability of the circuit to handle large signals. The linearity is typically measured in terms of the IIP3. The IIP3 of the amplifier is equal to the IIP3 of the degenerated transistor multiplied by 2 because of the potential divider at the input across the source impedance Z_0. Neglecting the effect of the non linearity of the transistor parasitic capacitor, the amplifier IIP3 is given by Lee:[5]

$$V_{IIP3} = 4 \cdot \sqrt{2 \cdot V_T} \cdot \left(1 + \left(\frac{I_C \omega L_E}{V_T}\right)^2\right)^{\frac{3}{4}} \tag{3.12}$$

This means that, unlike the noise factor, IIP3 gets better with frequency, and to obtain a wideband operation, special care should be taken at the lower end of the band.

3.4.2 Wideband Folded Cascode Amplifier Topology

In the proposed wideband design shown in Fig. 3.18, as in previous circuits a fourth-order doubly terminated band-pass filter is used to resonate the active part of the input impedance over the whole band. The typical cascode has been divided into two branches. The L_{T1} and L_{T2} have been added in order to bias the

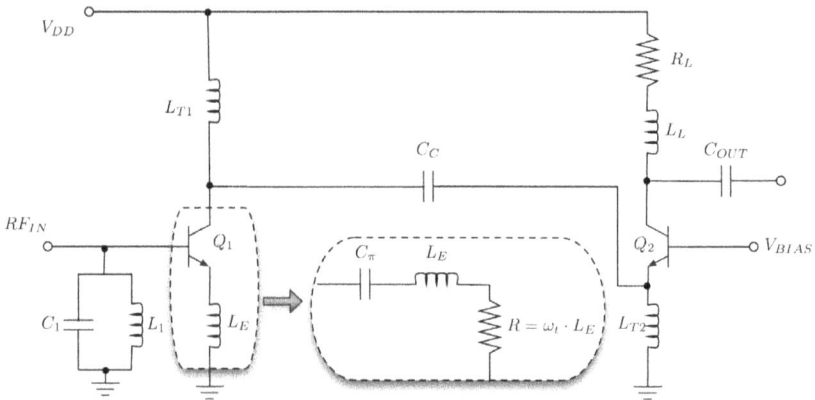

FIGURE 3.18 Simplified schematic of the LNA with wideband impedance matching.

collector of Q_1 and the emitter of Q_2, respectively. On the other hand, the capacitor C_C has been added in order to decouple the signal between the two branches. The output load is formed with a shunt peaking as in previous designs.

3.4.3 Experimental Results

To verify the functionality of the proposed low-voltage topology, a comparison is made between two UWB LNAs: (a) the conventional cascode topology and (b) the folded topology. The circuit was fabricated in the same 0.35 μm BiCMOS process as the conventional cascode amplifier. The die photographs of those circuits are shown in Fig. 3.19. As can be seen, thanks to the use of MLS inductors for L_{T1} and L_{T2}, the total chip size of both circuits is the same (665 × 665 μm). For our comparison, we have ensured that both designs were similar except for the use of the capacitively coupled resonating element in the low-voltage topology; that is, the same transistor geometries and the same biasing conditions were used in both designs.

Figures 3.20–3.23 show the measurements of both LNAs. Note that the measurements include the probe pads and buffer. This worsens the performance in comparison with the typical applications. In most of the wireless transceivers, the following stage of the LNA is a mixer, which is a capacitive load rather than a 50 Ω load.

Figure 3.20 shows the S_{21} and S_{11} measurements of both LNAs. As expected, the response of the two designs is

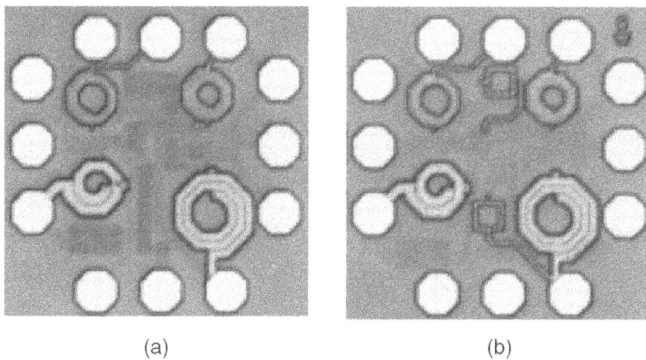

(a) (b)

Figure 3.19 (a) Cascode LNA and (b) folded cascode photograph.

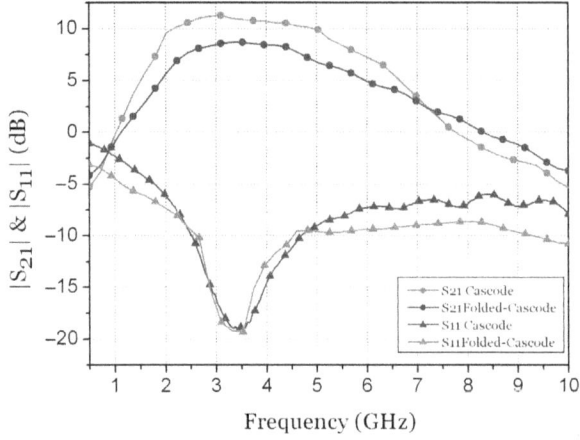

FIGURE 3.20 Measured S_{21} and S_{11} versus frequency.

approximately the same. As shown in Fig. 3.21, the folded cascode shows an enhanced NF with respect to the cascode LNA. This is due to the Q_2 shot noise filtering associated with the capacitively coupled resonating elements and the fact that the gain at Q_1 collector is greater than 0 dB because of the RF tanks.[28]

The measured IIP3 for the cascode and folded cascode LNAs is shown in Figs. 3.22 and 3.23, respectively. As it can be

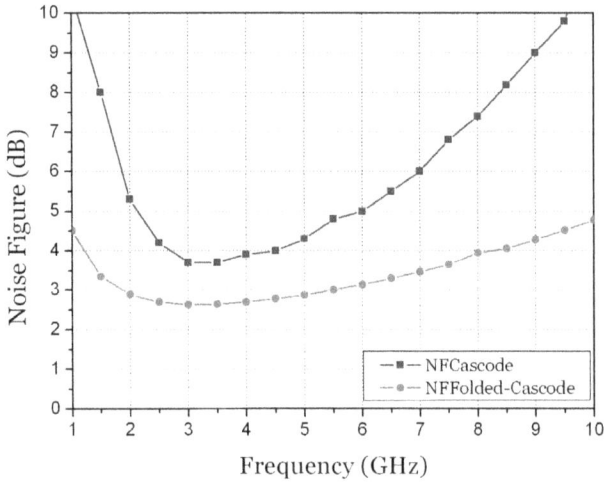

FIGURE 3.21 Measured noise figure versus frequency.

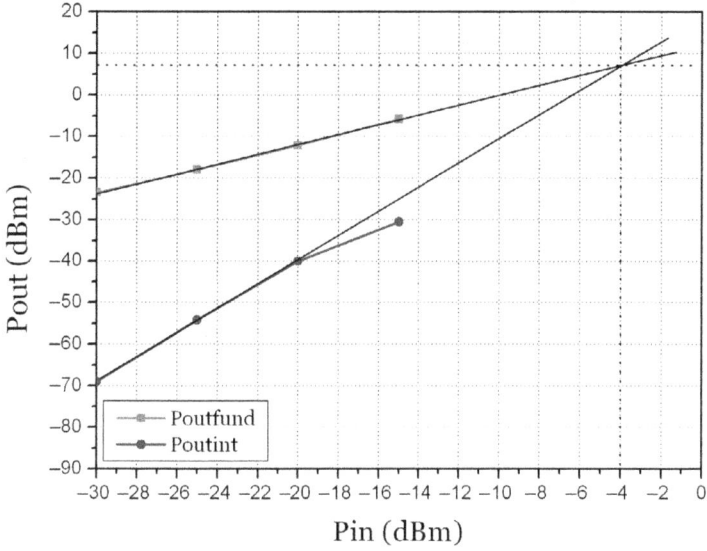

FIGURE 3.22 Measured IIP3 for cascode LNA.

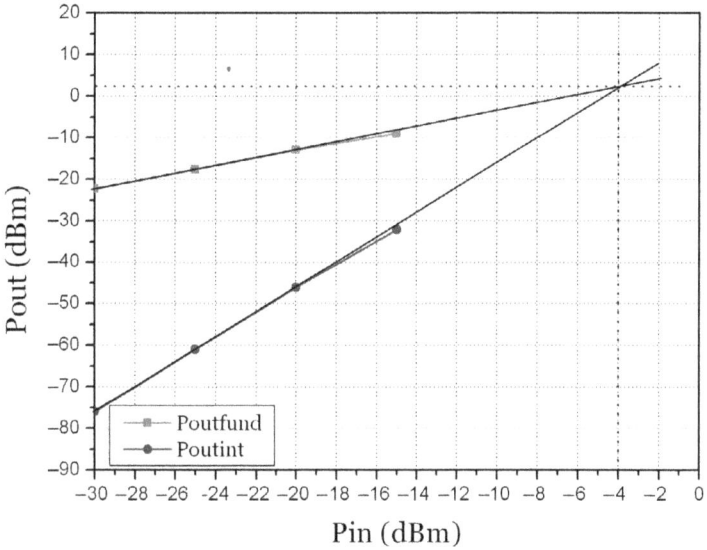

FIGURE 3.23 Measured IIP3 for folded cascode LNA.

Parameter	Cascode	Folded cascode
Supply voltage (V)	3	1.5
Power (mW)	18.93	18.87
$V_{CE}(Q_1, Q_2)$ (V)	1.5	1.5
$V_{BE}(Q_1, Q_2)$ (V)	0.8	0.8
NF @ 4 GHz (dB)	3.19	2.96
Gain-S_{21} @ 4 GHz (dB)	10.1	7.8
IIP3 @ 5GHz (dBm)	−4	−4

TABLE 3.2   Comparison between cascode and folded cascode LNA.

seen, the third-order input intermodulation point was determined to be equal for both designs.

In Table 3.2, a summary of the two designs is given. There is no significant degradation in performance when using the low-voltage cascode.

3.5 Conclusions

In this chapter three different alternatives for UWB low-noise amplifier have been presented. The first alternative, the wideband amplifier, presents a good frequency response but with a low gain flatness. To solve the gain flatness the modified shunt peaking amplifier was introduced. Finally the folded cascode topology has a low voltage operation and high linearity. Table 3.3 shows a performance summary of these circuits.

In comparison with the distributed amplifier the circuits developed in this chapter improve the power and the area consumption. To reduce area and power consumption, in the next chapter the feedback techniques are explored.

Design	Gain (dB)	BW (GHz)	NF (dB)	IIP3 (dBm)	Area (mm²)	P_{DC} (mW)
Wideband amplifier	12.5	5.3	4.3	−4	0.13	32
Modified shunt peaking	11.2	5	5	−4	0.29	56.1
Folded cascode	7.8	2.96	3	−4	0.13	18.93

TABLE 3.3   Wideband amplifiers specifications.

CHAPTER 4

Feedback Wideband Low-Noise Amplifiers

4.1 Introduction

In general, feedback techniques help to improve the amplifier's performance. The feedback loop helps to increase the gain and the bandwidth and, in some cases, even reduces the noise figure and power consumption. In this chapter, feedback techniques are explored in order to develop a wideband low-noise amplifier.

4.2 Circuit Analysis

Figure 4.1 shows a common emitter with shunt feedback. Ignoring the transistor capacitances, the voltage gain is given by

$$A_V = \frac{v_o}{v_i} = \frac{\frac{R_L}{R_F} - g_m R_L}{1 + \frac{R_L}{R_F}} \approx \frac{-g_m R_L}{1 + \frac{R_L}{R_F}} \tag{4.1}$$

where g_m is the transconductance of the transistor Q_1. This means that the gain without feedback $(-g_m R_L)$ is reduced by the presence of feedback.

The input impedance of the feedback amplifier also changes with respect to the open loop amplifier. Ignoring the base-emitter capacitance, the input impedance can be given by

$$Z_{in} = \frac{R_F + R_L}{(1 + g_m R_L)} \approx \frac{R_F + R_L}{g_m R_L} \tag{4.2}$$

Feedback results in a reduction of the role the transistor plays in determining the gain and therefore improves linearity. However, the presence of R_F may degrade the noise performance depending on the value of this resistor. By performing

69

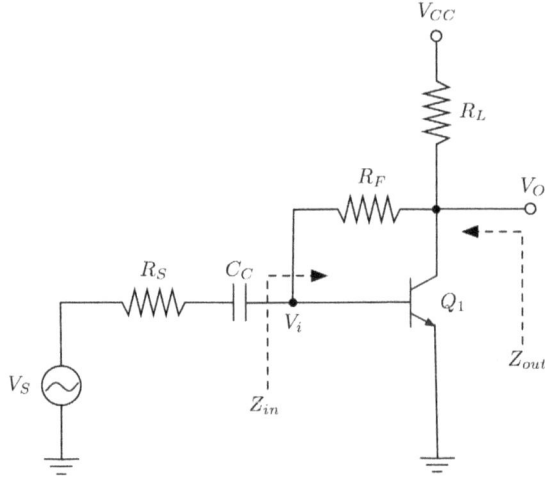

FIGURE 4.1 Schematic of the amplifier with shunt feedback.

the noise analysis, a simplified expression for the amplifier's noise factor is derived:

$$F = 1 + \frac{r_b + r_e}{R_S} + \frac{1}{2\, g_m R_S} + \frac{g_m R_s}{2\beta} + \frac{g_m R_S}{2\beta^2} + \frac{1}{2\, g_m} \frac{R_S}{R_F{}^2} + \frac{R_S}{R_F}$$

(4.3)

where r_b and r_e are the base and emitter parasitic resistances, and β is the small signal current gain. The noise analysis shows that the feedback resistor R_F can significantly affect the overall amplifier noise figure due to its relative magnitude with respect to the source resistance, R_S. The required linearity, typically measured in terms of the IIP3, is specified by:

$$IIP3_{LNA} \propto g_m{}^2 \propto I_{bias}{}^2$$

(4.4)

Intuitively, the higher the g_m, the larger the loop gain, which improves linearity.[29] A larger g_m means more current consumption. However, for high-frequency operation, more current consumption is usually needed to drive the parasitic capacitances and obtain enough gain. This results in a little flexibility in the choice of g_m. The voltage gain given by Eq. 4.1 sets a relation between R_L and R_F for a given g_m. As a result, the noise factor and input resistance are coupled because, as shown in Eqs. 4.2

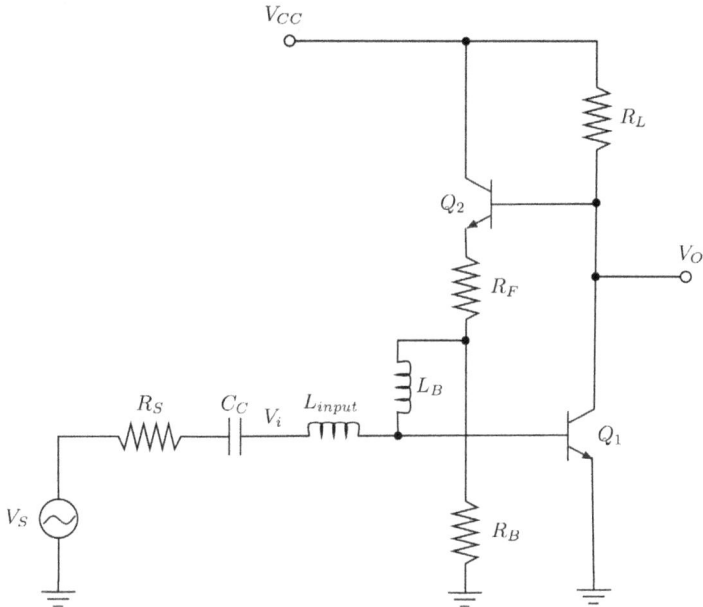

FIGURE 4.2 Schematic of the amplifier with active feedback.

and 4.3, they both depend on R_L and R_F. Because of this trade-off, it is generally difficult to achieve an arbitrarily low noise factor for an input impedance of 50 Ω with a reasonable current consumption.

To resolve this issue, the resistive feedback can be replaced by a feedback through an emitter follower. Figure 4.2 shows the proposed topology. The amplifier consists of a single stage in common emitter configuration and an emitter follower in the feedback path.

At low frequency the input impedance can be given by

$$Z_{in} = \frac{1 + g_{m2} R_F}{g_{m2}(1 + g_{m1} R_L)} \approx \frac{R_F}{g_{m1} R_L} \tag{4.5}$$

where g_{m2} is the transconductance of the transistor Q_2 in the source follower.

Under the input matching condition ($Z_{in} = 50\,\Omega$) and for the same voltage gain, the required value of R_F is now enhanced compared with the previous case, and from Eq. 4.3 a low noise factor is achieved.

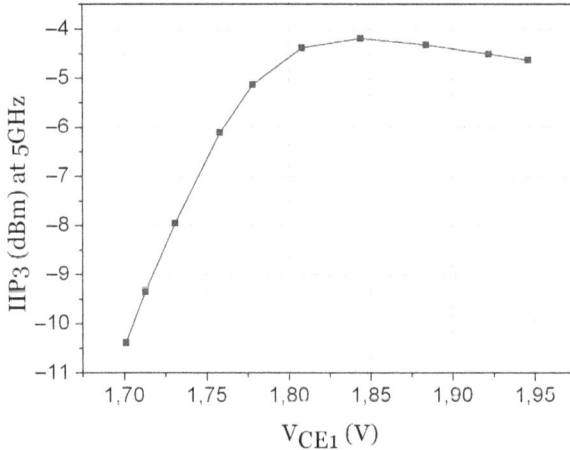

FIGURE **4.3**   IIP3 @ 5 GHz vs. V_{CE1} simulation.

Another advantage of the proposed topology is that, thanks to the use of R_F, R_B, and Q_2 in the feedback path, the collector-emitter voltage of Q_1 can be modified:

$$V_{CE1} \approx V_{BE2} + V_{BE1} \cdot \left(\frac{R_F}{R_B} + 1 \right) \qquad (4.6)$$

As a result, a higher f_T of the transistor and an improved large signal behavior can be achieved. This can be seen on Fig. 4.3, where the IIP3 is plotted as a function of V_{CE1}. As the figure suggests, there exists an optimum collector-emitter voltage to maximize the IIP3.

To enhance the bandwidth of an amplifier, such as in the previous chapter the inductor shunt peaking technique has been used. As stated in the previous chapter, this technique consists of adding an inductor in series with the load resistor to resonate out the capacitive parasitics and extend the circuit bandwidth. However, the inductance value of the added inductor can be large, and this consumes much chip area.

An inductor L_B placed inside the feedback loop is proposed. Figure 4.4 shows the simulation results with several L_B values. It clearly shows that the bandwidth of the amplifier is increased when increasing L_B. However, large values for the inductance of L_B lead to peaking in the frequency response.

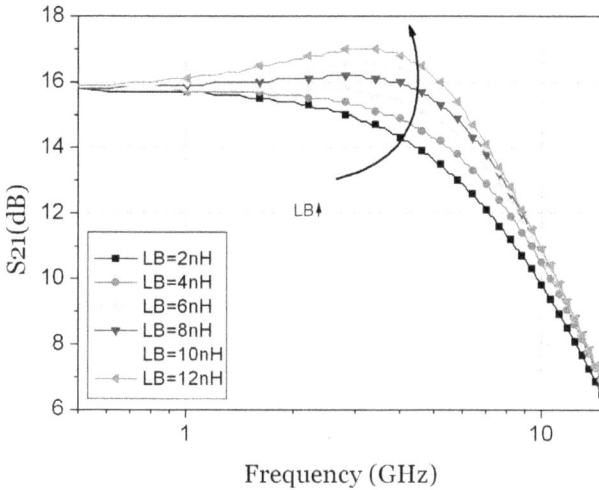

Figure 4.4 S_{21} vs. L_B simulation.

While this allows extending the bandwidth to higher frequencies, excessive peaking is undesirable for broadband communications systems that require a flat group delay. The quality factor of this inductor is only of minor importance in this application, due to its low noise contribution in the signal path. Indeed, the lower the quality factor, the higher the series resistance associated with it. This series resistance adds to R_F, enhancing the circuit noise performance.

In this chapter, we propose two options to implement L_B: a conventional spiral inductor and a modified miniature 3D inductor. A detailed study of the latter is presented in the next section.

The inductor L_{input} is used to achieve a good response over the whole frequency range (from 0 to 15 GHz). This coil is in the direct path of the signal but its inductance value is low. Fortunately, because low inductance spirals are achieved with a small number of turns, the quality factor is high, and in consequence, its contribution to the total noise figure will be low.

4.3 Modified Miniatured 3D Inductor

As explained earlier, the conventional approach to designing an integrated inductor on silicon is to layout a simple

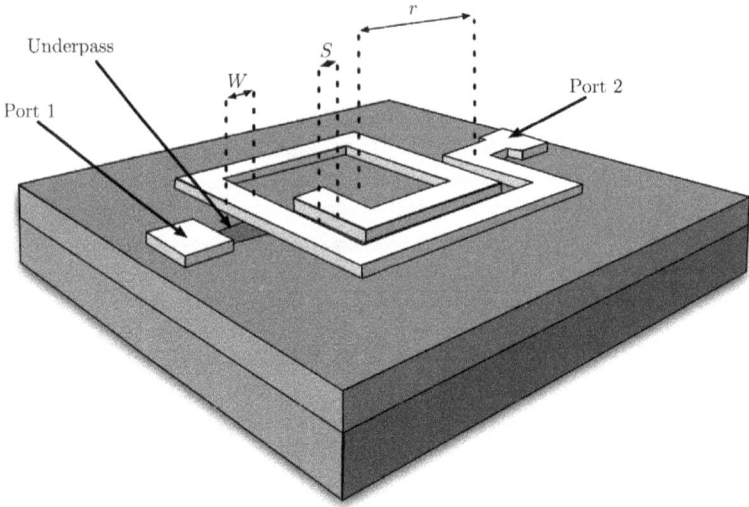

FIGURE 4.5 Layout and geometric parameter of an on-chip inductor.

metallic spiral directly on the substrate (see Fig. 4.5). At least two metal levels must be available, because an underpass is required to give access to one of the inductor's port. The challenge is to choose, for a given technology with fixed metal properties, the optimum combination of the number of turns (n), the metal width (w), the spacing between tracks (s), and the external radio (r) to provide a specific inductance and optimum quality factor at the frequency we are working on. This task is disturbed at high frequencies by the eddy current effects in substrate and metal turns and by the skin effect in the metal conductor.

Large inductance values typically combine with large areas and small Q_S: an increase in the number of turns in the spiral coil or an increase of the coil radii results in an increased magnetic flux and thus a higher inductance value, but also in a proportionally higher series resistance. This can be seen in Fig. 4.6, where the micro photograph and measured L and Q of a conventional high inductance spiral coil are shown (n = 3.5, w = 10 μm, s = 2 μm, and r = 120 μm). The low-cost employed technology, SiGe 0.35 μm, provides four metal levels. Three of them are similar, with equal thickness and conductivity, and the top level metal is thicker and more conductive. Standard coils are designed using this top metal, which presents a lower

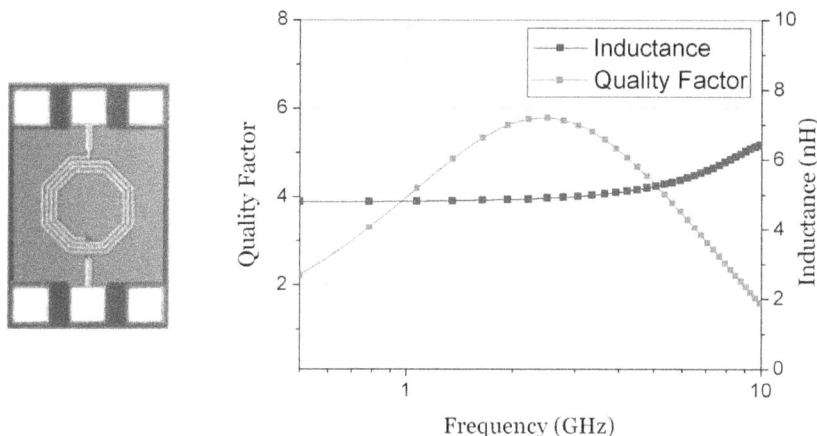

FIGURE 4.6 Microphotograph and measured quality factor and inductance of the proposed inductor.

series resistance and is far enough from the substrate to work at high frequencies.

As discussed in section 2.3.2, to save some silicon area, some authors employ stacked inductors. As can be seen in Fig. 2.6, it consists of series connected spiral inductors in different metal layers. Since the spirals are identical, the inductance value of each spiral separately will be the same. Spiral inductor segments in different layers, close to each other, have positive mutual inductance between them because current flows to the same direction. So, the total inductance value of the structure will increase due to the strong mutual coupling between the segments, and high inductance can be achieved in a small area.

However, the use of more metal layers amplifies the capacitive parasitic effects because of the metal-to-metal new capacitance and the increase of the metal-to-substrate oxide capacitance. Thus, the stacked inductor suffers from low self-resonance frequency. In an attempt to preserve the advantages of stacked inductor, and at the same time, to increase the resonant frequency and the quality factor, Tang et al. proposed in 2002 the miniature 3D inductors.[30] This structure consists of at least two or more stacked inductors by series connections, and every stacked inductor has only one turn in every metal layer.

For example, if there are two stacked inductors with different diameters, and one of them is a one-turn stacked inductor

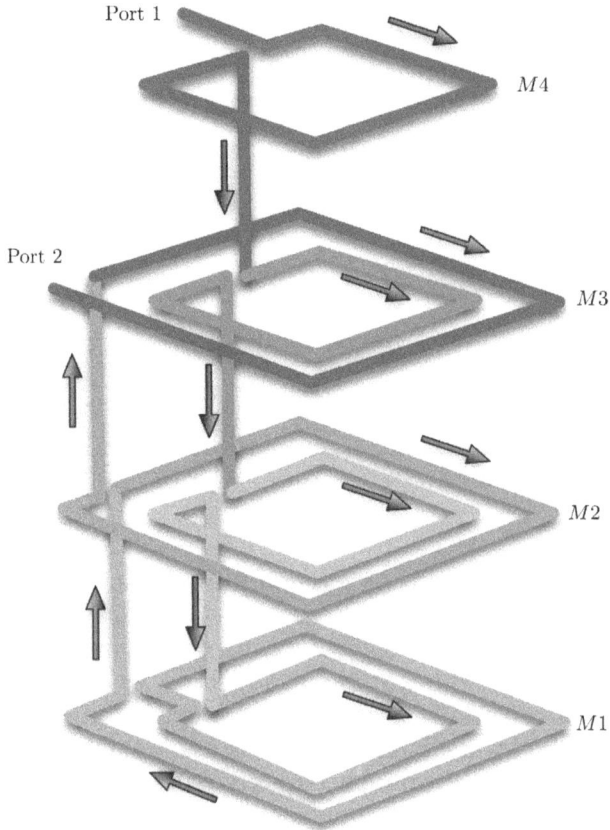

FIGURE 4.7 Miniature 3D inductor.

from the metal layer 4 to the metal layer 1, and the other is a one-turn stacked inductor from the metal layer 1 to the metal layer 3, then the miniature 3D inductor is formed by connecting two stacked inductors at the metal 1 layer, as Fig. 4.7 shows.

The proposed structure for inductor L_B consists of two 3D rectangular coils serially connected through the lower metal level, as shown in Fig. 4.8. This way, part of the magnetic flux generated by the coils is shared. Consequently, the structure total inductance is greater than the addition of both 3D inductance values separately.

Figure 4.9 (a) shows a micro photograph of the implemented inductor. The structure occupies an area of $98 \times 98 \ \mu\text{m}^2$, which corresponds to the 25% of the area occupied by a standard

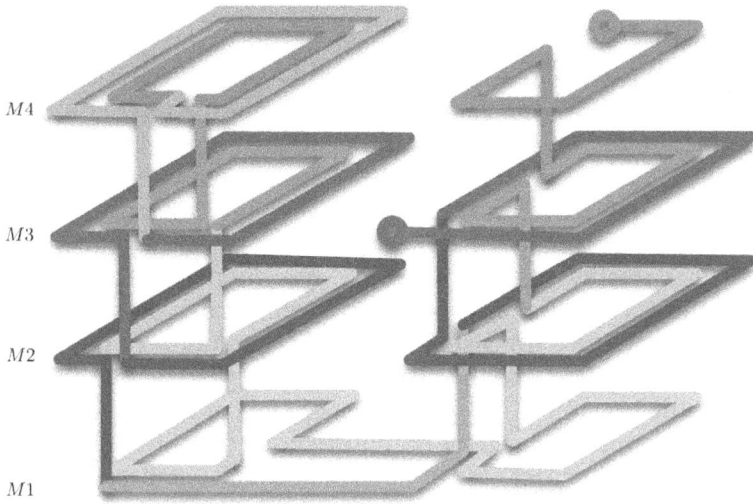

FIGURE 4.8 Modified miniature 3D inductor.

inductor with similar inductance response versus frequency. Figure 4.9 (b) shows the measured quality factor and inductance value of the proposed inductor. With this structure, in addition to reducing the occupied area a larger inductance is obtained.

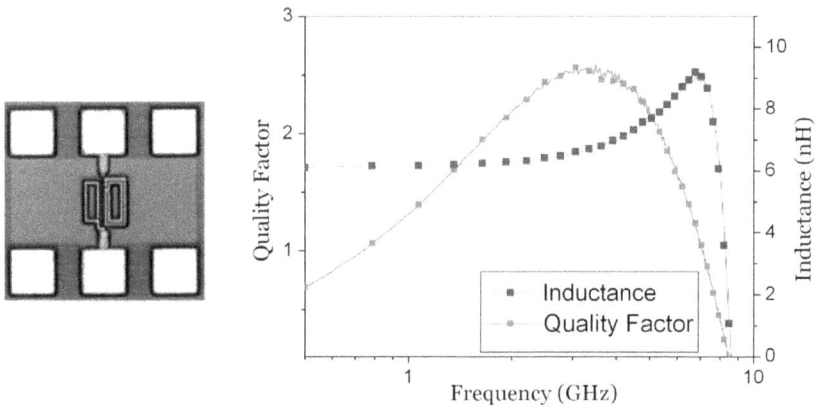

FIGURE 4.9 Microphotograph and measured quality factor and inductance of the proposed modified 3D inductor.

4.4 Circuit Design

To test the proposed technique, two LNAs, LNA1 and LNA2 were designed using conventional and modified miniature 3D inductors for L_B. Except for the structure employed to implement L_B, both circuits follow the same design considerations. The LNAs were fabricated using AMS SiGe 0.35 μm BiCMOS technology.

It is readily seen from Eq. 4.3 that the noise factor of the amplifier is determined by the collector and base shot noise of the first-stage transistor Q_1, the thermal noise of the shunt feedback resistor, and the thermal noise of Q_1's base and emitter parasitic resistors. The bias current of Q_1 is optimized together with its emitter length for minimum noise. In our final design, the effective emitter area of Q_1 is 36 μm^2. The emitter follower Q_2 contributes only little to the output noise and an effective emitter area of 1.6 μm is chosen.

The feedback resistor, R_F, has a large impact on the noise factor. To reduce its noise contribution, a large value is used in conjunction with the consideration for input impedance match. The choice of the feedback resistor determines the operating bandwidth. Low R_F increases the operating bandwidth while sacrificing the gain and noise performance. The use of the peaking inductor L_B in the feedback path alleviates this tradeoff, and its inductance value should be chosen as high as possible up to where it can meet the bandwidth requirement without excessive peaking. As explained in previous sections, the quality factor of inductor L_B is not relevant. In this case, achieving a high-value inductance in a small area is the major requirement. For that reason a modified miniature 3D inductor could be a better solution for L_B. A different situation is observed for L_{input}. This inductor is in series with the input and is used to help match the input impedance within the entire band. Its quality factor should be as high as possible, since its value affects the overall noise performance. Because the required inductance value is low, a conventional spiral inductor can be used in the implementation (1.5 turns).

The load, consisting of a poly-silicon resistor R_L, is designed to achieve a flat broadband gain over the entire UWB band. According to the analysis shown in previous sections, amplifier linearity depends on V_{CE1} selection through the feedback circuit. The value of R_B is chosen to maximize IIP3 and is calculated using bias current and supply voltage to bias the device at its low tolerance point to the distortion.

4.5 Experimental Results

The die photographs of LNA1 and LNA2 are shown in Fig. 4.10. The chip area excluding the test pads is $490 \times 355\ \mu\mathrm{m}^2$ for LNA1 and $330 \times 310\ \mu\mathrm{m}^2$ for LNA2. Note that the proposed technique achieves a 40% reduction of area, and as we show later, with minor performance degradation over the same circuit implemented with conventional inductors.

The S parameters and the noise figure of both circuits were measured using ground-signal-ground microwave probes. Both circuits operate with a supply voltage of 3.3 V and consume 4 mA.

Figures 4.11 and 4.12 show the measured gain and noise figure for 50 Ω source and load impedance. Both amplifiers provide a gain that varies from 14 to 7 dB in the band between 3.1 and 10.6 GHz, being greater than 1 dB from 0.1 to 15 GHz. The gain response is flat, which indicates that no

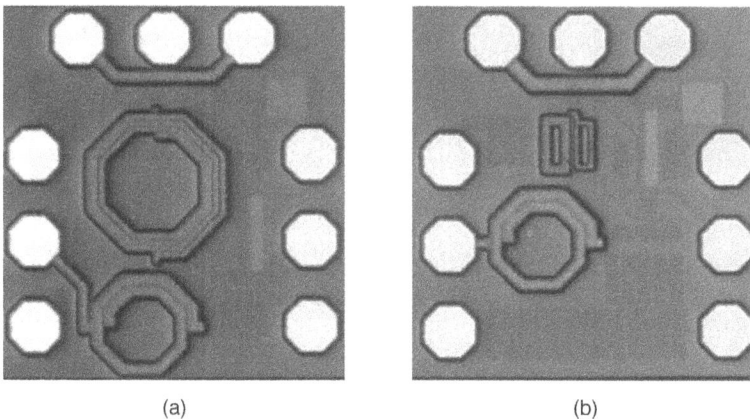

(a)   (b)

FIGURE 4.10   Chip photograph of (a) LNA1: with conventional spiral inductor and (b) LNA2: with modified miniature 3D inductor.

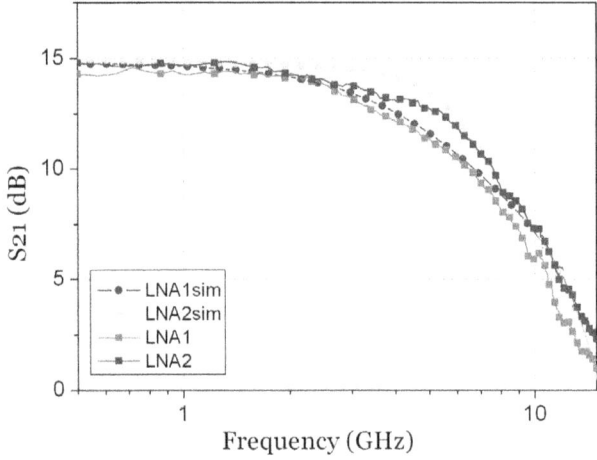

Figure 4.11 Simulated and measured gain.

excessive peaking was employed to obtain the desired band-width. The low-frequency gain of LNA1 is 15 dB, and the 3 dB bandwidth is 5.5 GHz. For LNA2, the gain is similar but the bandwidth is higher (6.7 GHz). This is due to the greater L_B

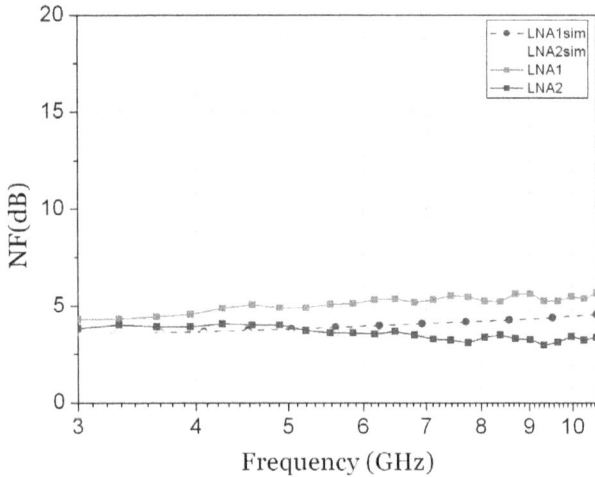

Figure 4.12 Simulated and measured noise figure.

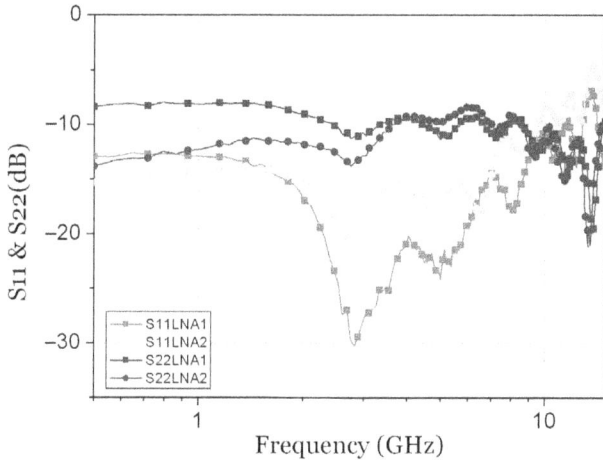

FIGURE 4.13 Measured input and output return loss (S_{11} and S_{22}).

inductance value obtained with the 3D inductor (6 nH) compared with the conventional spiral inductor (5 nH). Another positive effect associated with the use of a 3D inductor in the feedback path is improved noise performance. As shown in Fig. 4.12, the measured noise figure of LNA1 is 4.2 dB for low frequencies and rises to 5.6 at 10.6 GHz. The noise figure of LNA2 is between 2.9 and 4 dB from 3.1 to 10.6 GHz. The noise figure improves at high frequencies due to the added resistance associated with the low Q 3D inductor.

Figure 4.13 shows the measured input and output return loss of both amplifiers. S_{11} and S_{22} for LNA2 are lower than −9 dB between 3.1 and 10.6 GHz.

The two-tone test for IIP3 is shown in Fig. 4.14 for LNA1 and LNA2. The test is performed at 5 GHz. Tone spacing is 100 KHz. LNA1 and LNA2 achieve an IIP3 of −3.4 dBm and −4.4 dBm, respectively.

The measurement results of the two wideband LNAs and several previously published results are listed for comparison in Table 4.1. As shown in Table 4.1, the modified 3D inductor active feedback LNA achieves low noise figure, high gain, and high IIP3 while simultaneously occupying the lowest area yet published in a commercial SiGe BiCMOS process.

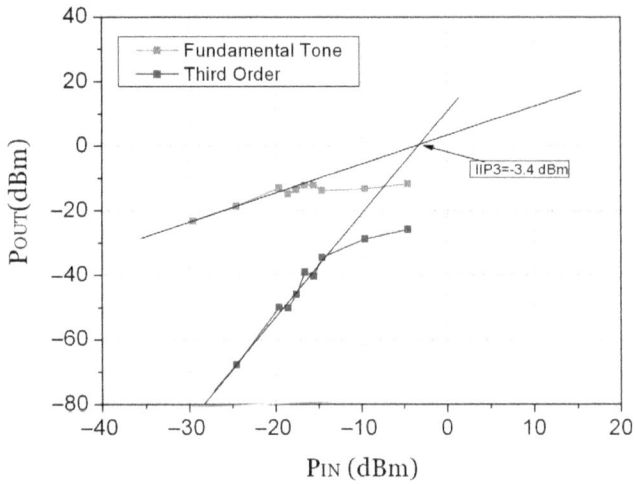

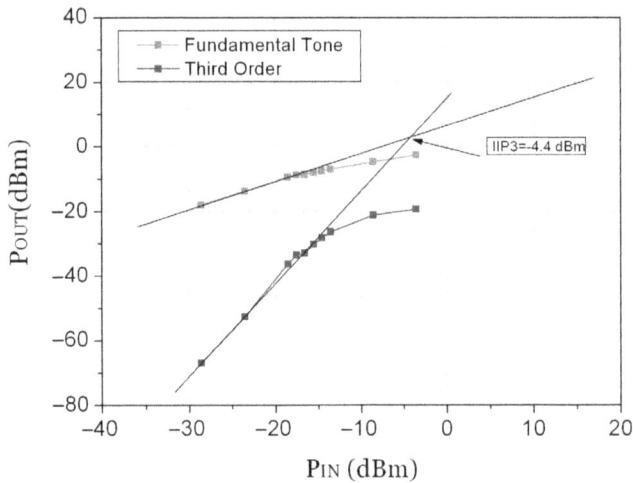

FIGURE 4.14 Measured two-tone test at 5 GHz LNA1 (a) and LNA2 (b).

Ref.	S_{21} (dB)	NF (dB)	3dB BW (GHz)	IIP3 (dBm)	P_{dc} (mW)	Active Area (mm)	Tech.
2	9.3	<9	2–23	−6.7	9	1.1	0.18 μm CMOS
31	21	<4.5	2–10	>−5.5	30	0.55	0.18 μm SiGe
32	9.3	<9.2	2.3–9.2	>−6.7	9	0.66	0.18 μm CMOS
33	8.5	<5.3	1.3–10.7	>8	4.5	1	0.18 μm CMOS
33	8.2	<5.5	1.3–12.3	>8	4.5	1	0.18 μm CMOS
15	10.6	<5.4	0.01–14	>10	52	1.35	0.18 μm CMOS
34	20	<4.5	3–10	>−11.75	42.5	0.18	0.18 μm CMOS
35	22	<3.9	3.1–14.5	>−32.5	13.2	0.49	0.18 μm SiGe
36	15.3	<2.98	3.1–10.6	>−8.5	9	0.87	0.25 μm SiGe
37	12	<4	2–10	>1.9	24	0.25	0.13 μm CMOS
38	13	<3.3	2–10	>−7.5	9.6	0.88	0.18 μm SiGe
38	11.5	<3.5	2–10	>−7.5	7.2	0.88	0.18 μm SiGe
39	11.5	4.7	3.1–10.6	−10	10.57	0.665	0.18 μm CMOS
Std.ind	14	<5.6	0.1–5.5	>−3.4	13.2	0.1	0.35 μm SiGe
3D ind.	14	<4	0.1–6.7	>−4.4	13.2	0.1	0.35 μm SiGe

TABLE **4.1** Comparative results.

4.6 Conclusions

In this chapter an alternative to implementing wideband low-noise amplifiers, a feedback low-noise amplifier was presented. To reduce the area, modified 3D inductors were used.

Table 4.2 shows a performance summary. In comparison with the developed circuits in previous chapters, the feedback

Design	Gain (dB)	BW (GHz)	NF (dB)	P_{1dB} (dBm)	Area (mm^2)	P_{DC} (mW)
Standard inductor	14	5.5	<4	−3.4	0.17	13.2
3D inductor	14	6.7	<4	−4.4	0.10	13.2

TABLE **4.2** Wideband feedback amplifiers specifications.

amplifiers have better performance with the minimum area and minimum power consumption.

In spite of the reduction in area achieved with the developed circuits, most part of the area is occupied by the integrated inductors. To find the maximum area reduction, the next chapter is devoted to exploring the inductorless techniques.

CHAPTER 5

Inductorless Techniques

5.1 Introduction

The number of inductors and the area occupied by them pose an inconvenience in radiofrequency integrated circuit design. In previous chapters, a reduction in area was achieved with different topologies of inductors, but the inductors were still necessary in these designs.

In order to reduce the area, it is important try to avoid the use of inductors; for this reason, this chapter is devoted to exploring inductorless techniques with the design of a frontend based on a common gate inductorless LNA and a quadrature Gilbert cell mixer.

5.2 Common Gate LNA

Common gate (CG) LNA is widely used in wireless communications.[1,5,40,41] In this section, the CG LNA is studied, that is the relationship between the input matching and voltage gain or the noise figure.

5.2.1 Input Matching and Voltage Gain

The desired input impedance of a CG input stage is achieved by adjusting the bias current, aspect ratio, and overdrive voltage. Thus, for an input impedance of 50 Ω, the objective is to obtain $1/g_m$ of approximately of 20 mS. The CG stage does not suffer from the Miller effect, and thus an adequate reverse isolation can be achieved with a single transistor stage. Therefore, the input matching network and load can be designed separately.

Large impedance toward the signal ground is needed to steer the signal into the input transistor source. This can be achieved with a current source I_{BIAS} shown in Fig. 5.1 (a). That

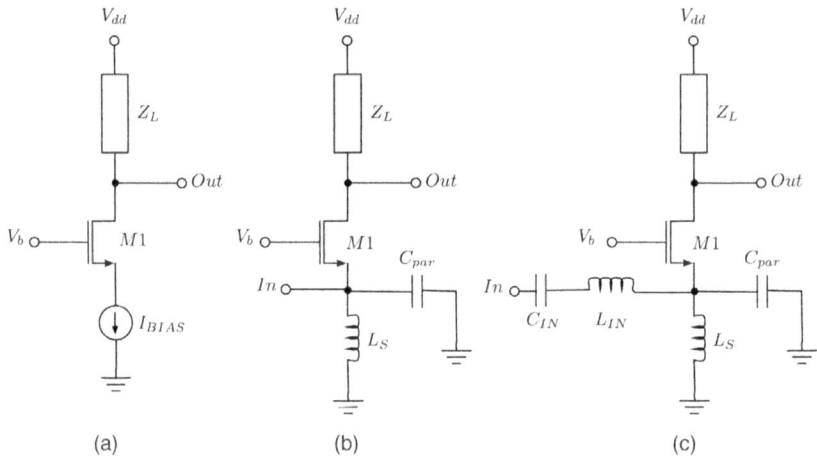

FIGURE 5.1 Common gate LNA input interfaces: (a) current source, (b) parallel LC resonator, and (c) series and parallel LC resonators.

topology is not typically utilized in LNAs since the current source I_{BIAS} increases the noise. A better noise performance is achieved by using a source inductor L_S as shown in Fig. 5.1 (b). The L_S forms a parallel LC resonator with the parasitic capacitance C_{par} associated with the source node of the M1. When on-wafer measurements are not applicable, the source node typically needs to be connected either to the package or printed circuit board (PCB) by using a bondwire inductance L_{IN} as shown in Fig. 5.1 (c). The L_{IN} is resonated at the desired frequency with a DC blocking capacitor C_{IN} that can be either an on-chip or an external component.

In Fig. 5.1 (b) and Fig. 5.1 (c), the capacitor C_{par} includes the parasitic capacitances at the source node, that is, source-body junction capacitance of M1, substrate capacitance of L_S, and capacitance caused by the bonding pads and on-chip metal wiring. Furthermore, the value of the source inductor L_S can be decreased by adding an additional shunt capacitor C_S in parallel with L_S (the C_S is not shown in Fig. 5.1 (b) and Fig. 5.1 (c)). Therefore, all the capacitance at the source node can be included in a single source capacitor C_T used in the following calculations:

$$C_T = C_{GS} + C_{PAR} + C_S \tag{5.1}$$

The input impedance Z_{IN} of a CG input stage, shown in Fig. 5.1 (b), can be calculated as

$$Z_{IN} = \frac{sL_S}{1 + sL_Sgm + s^2 + L_SC_T} \tag{5.2}$$

The source inductance L_S resonates with the capacitance C_T at the frequency of

$$\omega_o = \frac{1}{\sqrt{L_SC_T}} \tag{5.3}$$

and at that frequency, Eq. 5.2 simplifies to $1/g_m$. The impedance Z_{IN} of a CG input stage shown in Fig. 5.1 (c) is

$$Z_{IN} = \frac{1 + s^2C_{IN}L_{IN}}{sC_{IN}} + \frac{sL_S}{1 + sL_sgm + s^2L_SC_T} \tag{5.4}$$

With a perfect impedance matching, $1/gm = R_S$, where R_S is the source output resistance. The voltage gain of the CG amplifier becomes a division of output load versus the source impedance, that is, Z_L/R_S. The assumption is valid if the drain-to-source resistor r_{ds} is much larger than the load resistance at the drain. Otherwise, the gain and input resistance formulas should be modified to:

$$A_V = \frac{g_m Z_L}{\left(a + \frac{Z_L}{r_{ds}}\right)} \tag{5.5}$$

and

$$R_{IN} = \frac{1}{gm}\left(1 + \frac{Z_L}{r_{ds}}\right) \tag{5.6}$$

5.2.2 Noise of a CG Stage

The noise factor of a CG LNA is expressed as follows by Liao et al.:[42]

$$F = 1 + \frac{\gamma}{\alpha}\left(\frac{1}{1 + \chi}\right)^2 \frac{1}{gmR_S} \tag{5.7}$$

where γ is the coefficient of channel thermal noise, gm is the transistor transconductance, χ is the ratio of the transistor substrate transconductance gm_b and gm, R_S is the source resistance, and α is gm/gd_o. Because the minimum NF of a common-source LNA increases along with the frequency, CG LNAs can be a better option at very high frequencies. When χ is neglected and perfect input matching is assumed, the minimum noise factor typically presented in the literature is achieved:

$$F = 1 + \frac{\gamma}{\chi} = \frac{5}{3} = 2.2 \, dB \tag{5.8}$$

With imperfect input matching, the noise factor can be lowered according to

$$F = 1 + \gamma \frac{1 + S_{11}}{1 - S_{11}} \tag{5.9}$$

where α is neglected.

Equation 5.8 does not take into account the noise of the load. If resistive load R_L is used, taking into account its noise contribution and assuming that $gm = 1/R_S$, the noise factor becomes

$$F = 1 + \frac{1}{gm\,R_S} \left(\frac{\gamma}{\alpha} + \frac{(1 + gm\,R_s)^2}{gm\,R_L} \right) \simeq 1 + \frac{\gamma}{\alpha} + \frac{4R_S}{R_L} \tag{5.10}$$

Thus, the resistive load can make a significant contribution to the overall noise.

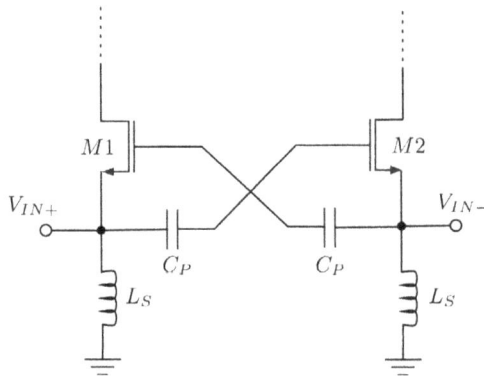

FIGURE 5.2 Capacitor cross-coupled CG stage.

5.2.3 Differential Operation of CG Stage

In Fig. 5.2, the capacitor cross-coupling method, which is suitable for differential input configurations, is presented. Due to the capacitor divider between C_{GS} and coupling capacitance C_p, the inverting gain is approximately $A = C_p/(C_p + C_{GS})$, which is always less than one.

5.3 Mixer Design

5.3.1 Quadrature Mixers

Normally, the down-conversion to zero or IF frequency is usually performed with quadrature (I/Q) mixers.[43] For I/Q mixers, there are different possibilities to implement the interface between input transconductor and switch quad. The first topology is shown in Fig. 5.3. The RF signal is fed into two separate input stages, which drive their respective switch quads. Another possibility is to utilize a single input stage, which drives both switch quads, as shown in Fig. 5.4.

Since the transconductor of the mixer, shown in Fig. 5.4, drives both switch quads, the conversion gain is 3 dB lower

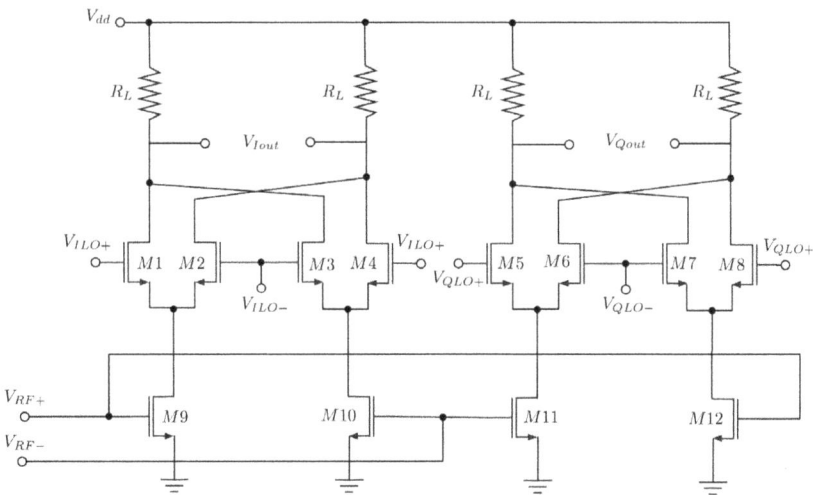

FIGURE 5.3 Two separate Gilbert mixers driven by quadrature LO signals.

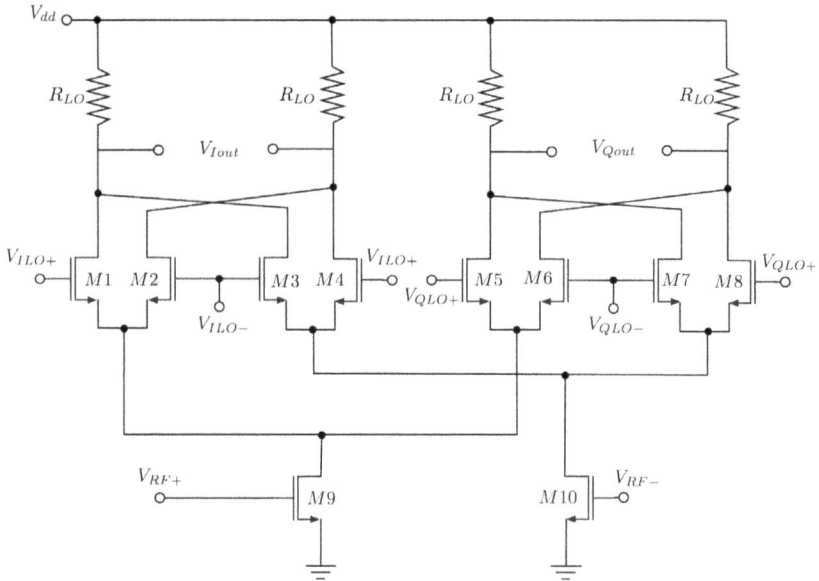

FIGURE 5.4 Quadrature mixer with a single input stage.

than the basic Gilbert cell mixer of Fig. 5.3. In adittion, complete switching requires larger LO amplitude when quadrature switch quads are driven from a single transconductor.

5.3.2 Mixers with Current Boosting

The current boosting method[40,43,44] is used in the mixer and it is shown in Fig. 5.5. The bias for the input and switching stages can be optimized separately with current boosting. For proper gain and linearity performance, the input transconductance stage should be biased with higher current. However, the performance of the switch quad may require quite a low current level for optimum operation. The conversion gain increases for two main reasons: the mixer requires a lower LO swing to switch completely, and a larger load resistor value can be used to increase the voltage gain. Alternatively, if the load resistor value is kept unchanged, mixer design for lower supply voltages is alleviated with current boosting, since the voltage drop at the resistive load is reduced.

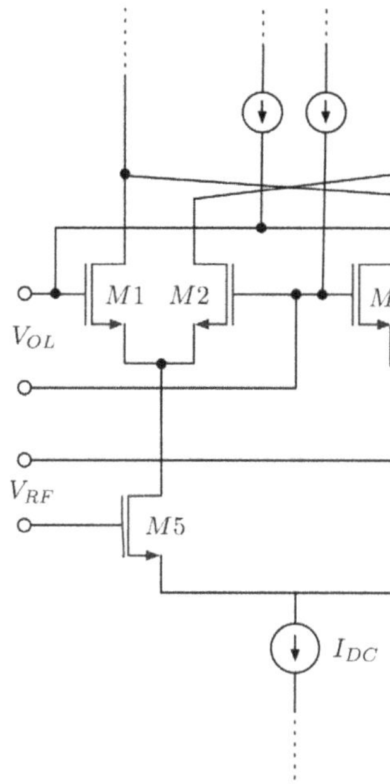

FIGURE 5.5 Current boosting with constant current source.

5.4 Inductorless Operation

Figure 5.6 shows a simplified interface between the LNA and mixer. Biasing and AC coupling have been omitted for clarity. In this figure, M_0 represents the LNA transconductor, wich is realized as a CG stage. Further, M_1 represents the RF transconductor of the mixer. The circuit in Fig. 5.6 (a) shows the use of conventional inductive peaking to extend the bandwidth at the LNA output. (It is really the LNA-mixer interface bandwidth that is being extended.) This bandwidth extension occurs due to the addition of a zero in the transfer function caused by the inductor. A zero can also be introduced by capacitive degeneration as shown in Fig. 5.6 (b). In this circuit, the R_C combination of C_S and R_S provides a zero that extends the high-frequency response. Effectively, capacitive degeneration provides bandwidth extension properties similar to inductive peaking.[45]

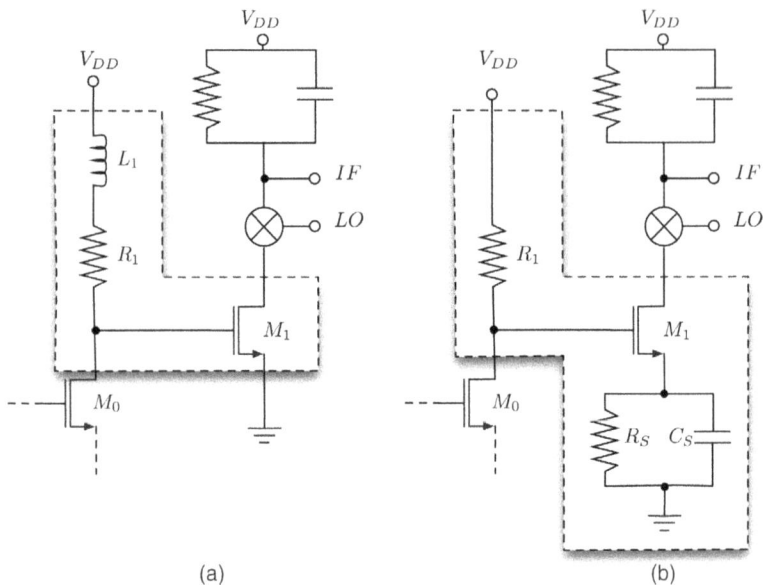

FIGURE 5.6 Simplified circuits for (a) inductive peaking and (b) capacitive peaking.

5.5 Experimental Results

5.5.1 Frontend I

Figure 5.7 shows an RF frontend composed by a differential common gate shunt peaking LNA followed by a differential double-balanced Gilbert mixer. The layout of the circuit, named as Frontend I, is shown in Fig. 5.8. The Frontend I chip area, including the test pads, is $1410.63 \times 693.39 \ \mu m^2$.

Conversion gain and noise figure simulation results are shown in Figs. 5.9 and 5.10, respectively, for a 400 MHz channel located at the center of the band. Frontend I has a conversion gain of 12.1 dB at 5.2 GHz and a noise figure of 11.2 dB (IF = 200 MHz).

The two-tone test for third-order intermodulation distortion of Frontend I is shown in Fig. 5.11. The test was performed at 5 GHz, and an IIP3 of -5.7 dBm was obtained.

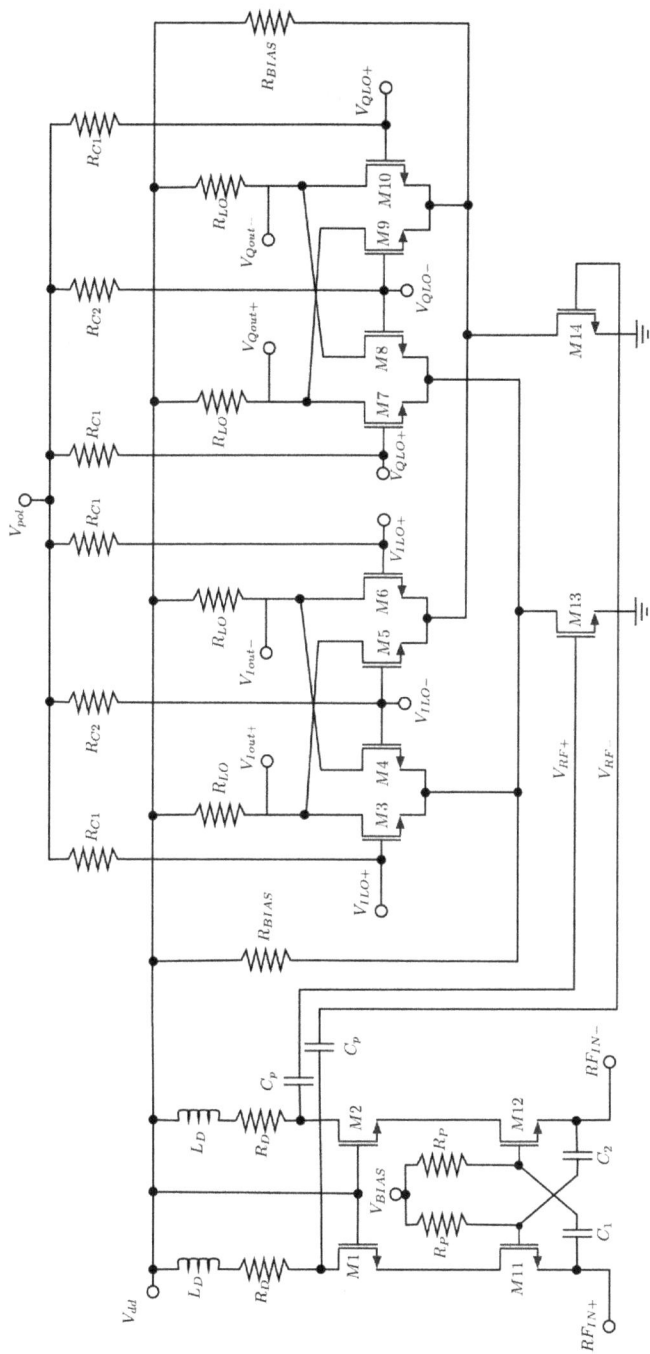

FIGURE 5.7 Frontend I schematic.

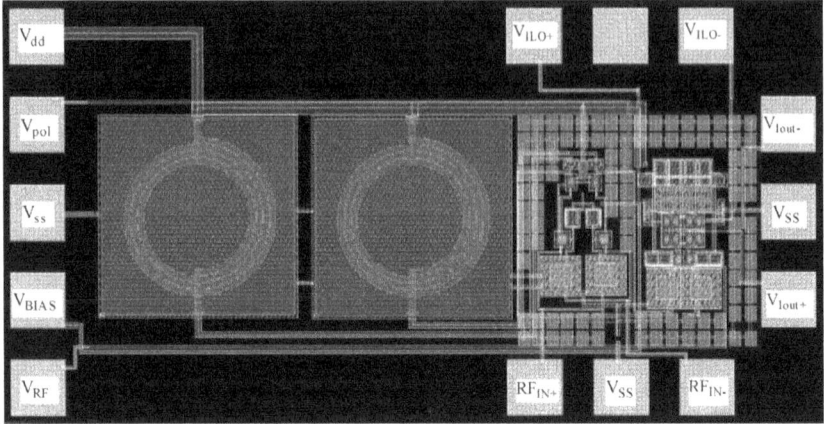

FIGURE 5.8 Frontend I layout.

5.5.2 Frontend II

Figure 5.12 shows an RF frontend composed of a differential common gate resistive load LNA followed by a differential double-balanced Gilbert mixer with capacitive degeneration. The layout of the circuit, named as Frontend II, is shown in Fig. 5.13.

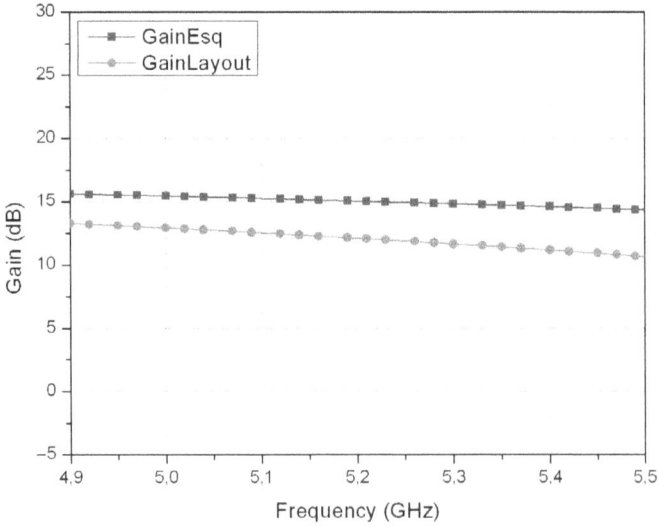

FIGURE 5.9 Frontend I conversion gain simulation results.

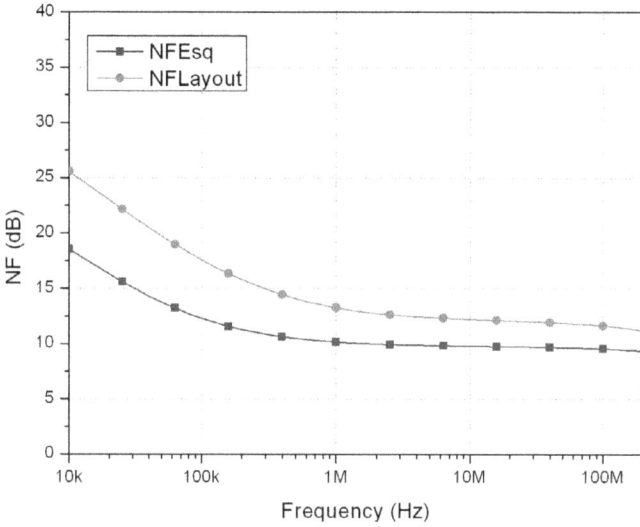

FIGURE 5.10 Frontend I noise figure simulation results.

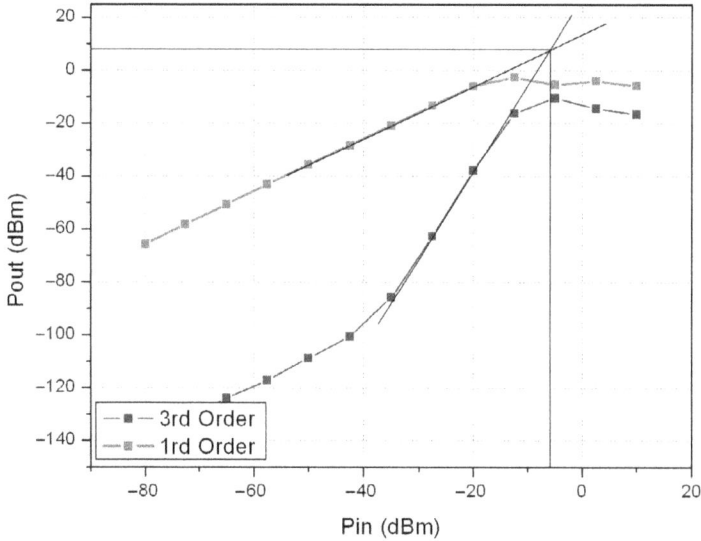

FIGURE 5.11 Frontend I IIP3 simulation results.

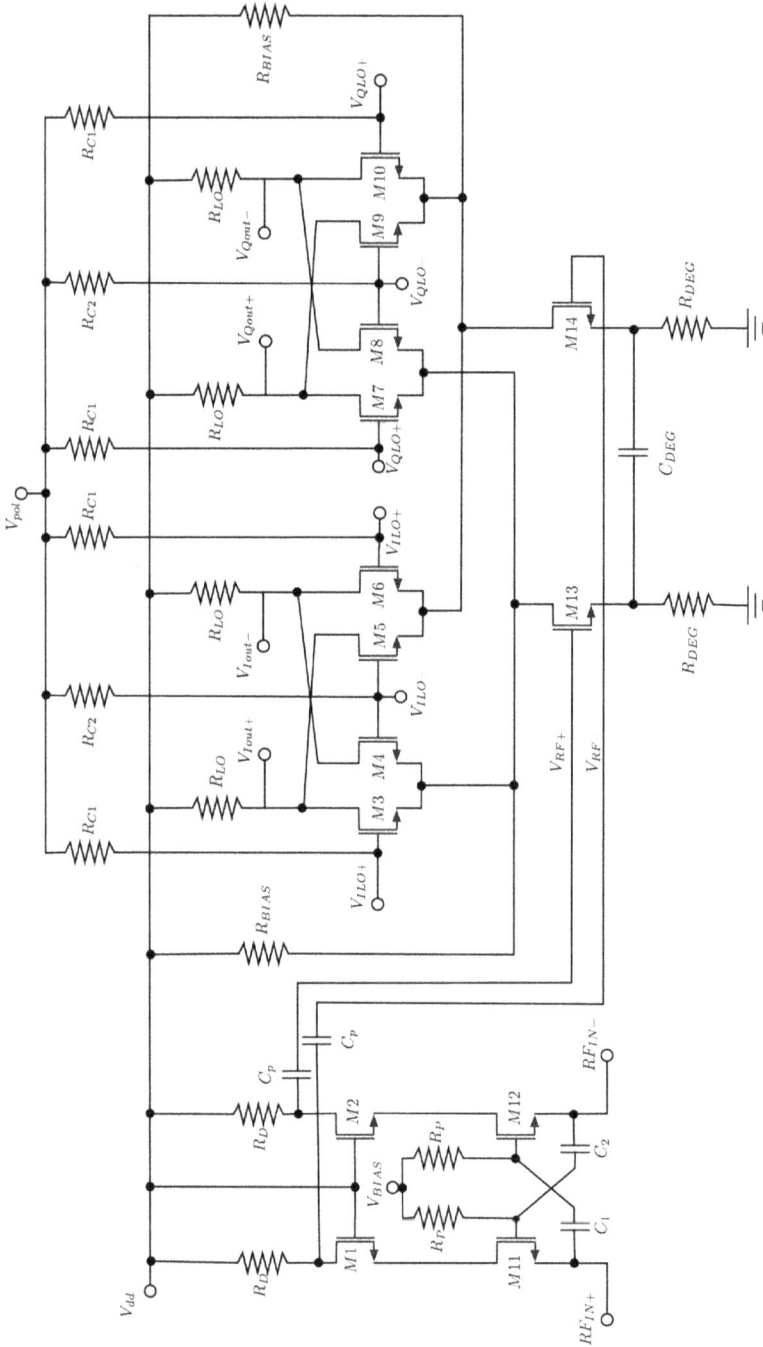

FIGURE 5.12 Frontend II schematic.

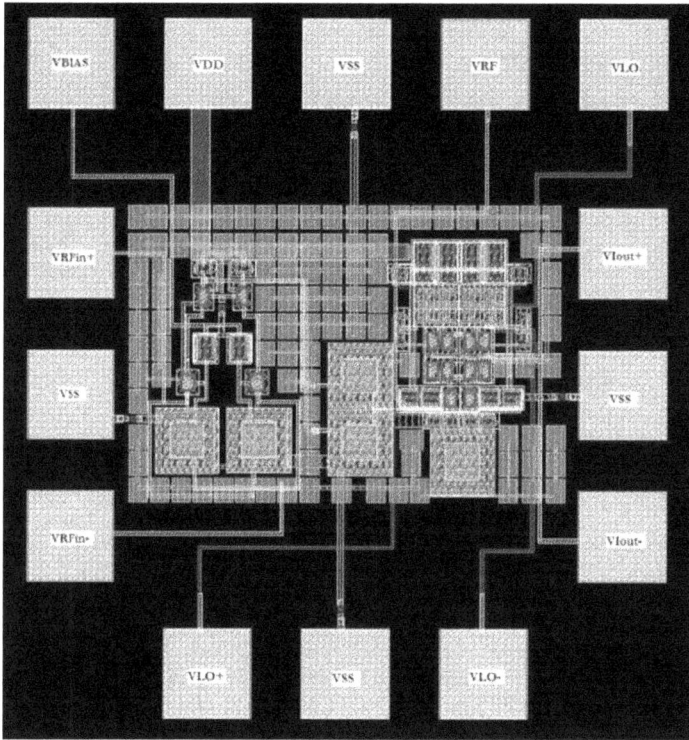

FIGURE 5.13 Frontend II layout.

The Frontend II chip area, including the test pads, is 698.89 × 744.76 μm^2. Due to the non-existence of inductors, the occupied area of this design is 54% smaller than Frontend I.

Conversion gain and noise figure simulation results are shown in Figs. 5.14 and 5.15, respectively, for a 400 MHz channel located at the center of the band. Frontend II has a conversion gain of 7.2 dB at 5.2 GHz and a noise figure of 13.71 dB (IF = 200 MHz).

The linearity of the Frontend II was evaluated with a two-tone test. The result is plotted in Fig. 5.16. The IIP3 is −2.1 dBm.

5.5.3 Comparison Between Frontends

The inductorless wideband amplifier can be designed if the amplifier works in conjunction with a Gilbert cell mixer because the mixer input is part of the low-noise amplifier output

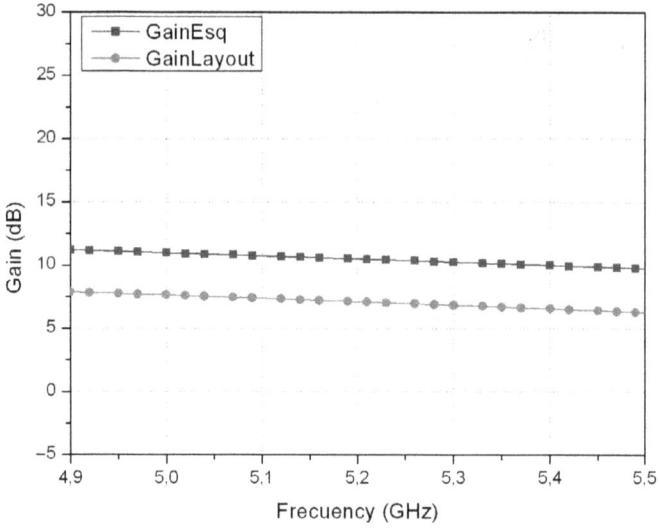

Figure 5.14 Frontend II gain simulation results.

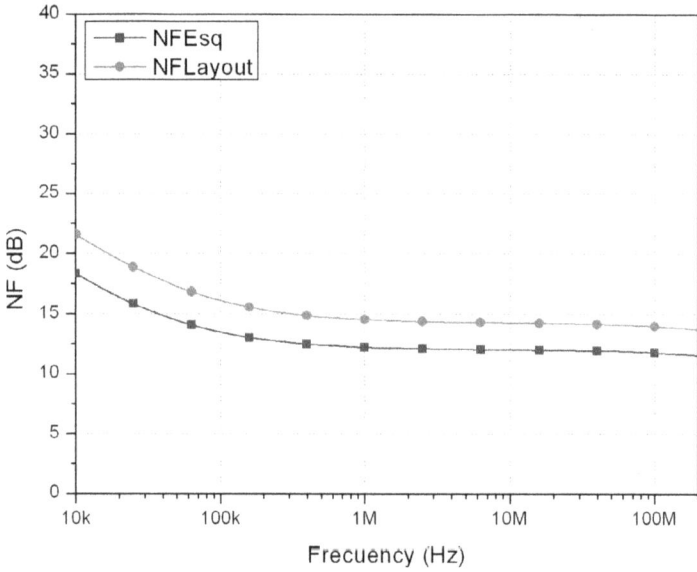

Figure 5.15 Frontend II noise figure simulation results.

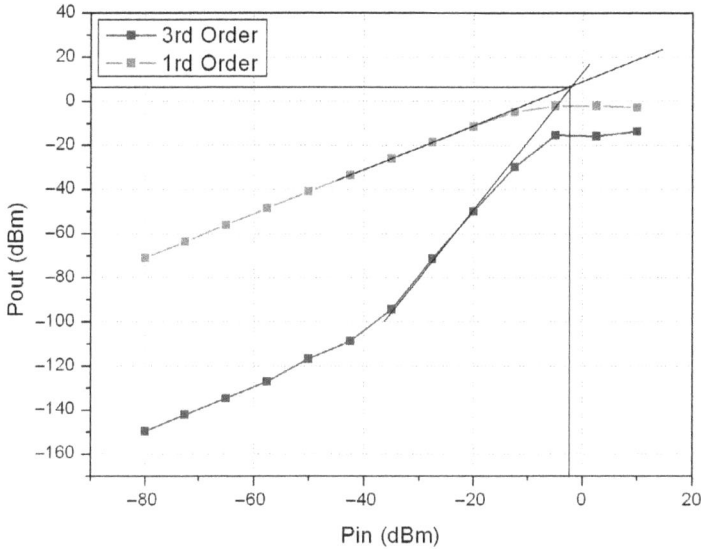

FIGURE 5.16 Frontend II IIP3 simulation results.

matching network. For this reason, in this chapter also a Gilbert cell mixer with current boosting has been designed.

To validate the reduction in area without a significant performance degradation, another wideband low-noise amplifier with shunt peaking followed by a Gilbert cell mixer was designed.

Table 5.1 shows the performance comparative of both frontends. Simulation results show that the proposed topology with the same bandwidth has better linearity and a comparable noise figure and uses less power. The silicon area for the inductorless LNA and I/Q mixers is 54% less than the traditional

Design	Frontend I	Frontend II
NF (dB)	11.2	13.7
Gain (dB)	12.1	7.2
IIP3 (dBm)	−5.6	−2.1
Consumption (mW)	16	14
Area (mm^2)	0.97	0.52

TABLE 5.1 Frontends performance summary.

inductor-based designs, thereby showing area savings and improved portability.

5.6 Conclusions

This book focuses on optimizing the area and power consumption of low-noise amplifiers for ultra-wideband communications. To have a reference set of specifications for the LNA, in Chapter 1 the ECMA-368/ISO/IEC 26907 standard was studied. From this study a reference system was designed and based on budget simulations, the LNA reference specifications were extracted.

In Chapter 2 the most classical type of wideband amplifier was presented, the distributed amplifier. With this topology a wideband operation is obtained with a relatively constant performance through the whole band. The main drawbacks of this topology are a high power consumption and a large area due to the significant number of inductors employed. In order to reduce the area, three different versions of the distributed amplifier were designed. The first one is a classical structure without any area-saving technique; it was used as a reference design. In the second version, a compacting technique is used. The objective here was to reduce the impact of mutual coupling between inductors through the correct coiling of the spirals. Finally, the third version employs multilevel stacked inductors achieving the most important area reduction (36%) with a minimum influence over the distributed amplifier performance.

Although an important saving in area was achieved, the distributed amplifier still uses a large area. For this reason, in Chapter 3 three alternative wideband amplifiers were studied. The first one is a classical cascode topology transformation to wideband operation. In this design, the input narrowband matching network is transformed into a wideband matching network and the output load is also transformed to a wideband load. The immediate consequence is a reduction in the area and power consumption, but the problem is that it is difficult to keep the gain constant over the whole band.

To solve the gain flatness problem, the modified shunt peaking topology was introduced in the same chapter. This technique improves the gain flatness but it increases the area due to the introduction of an additional inductor. Another alternative

Design	Gain (dB)	BW (GHz)	NF (dB)	IIP3 (dBm)	Area (mm²)	P_{DC} (mW)
Distributed amplifier 1	7	6.5	5	21.3	0.74	90
Distributed amplifier 2	7	6.5	4.5	21.4	0.61	90
Distributed amplifier 3	5.5	6.5	6	20.2	0.47	90
Wideband amplifier	12.5	3.6	4.3	−4	0.13	32
Modified shunt peaking	11.2	4	5	−4	0.29	56.1
Folded cascode	7.8	2.96	3	−4	0.13	18.93
Feedback amplifier standard ind.	14	5.6	<4	−3.4	0.17	13.2
Feedback amplifier 3D ind.	14	6.8	<4	−4.4	0.10	13.2
Receiver 1 (inductor based)	12.1	5	11.2	−5.6	0.97	16
Receiver 2 (inductorless)	7.2	5	13.7	−2.1	0.52	14

TABLE **5.2**   Circuit specifications.

studied in this chapter was a folder cascode topology. With this structure, a low-voltage operation is possible with a minimal influence over the circuit performance. In this design, stacked inductors were introduced as folding elements, achieving an important reduction in the area.

In Chapter 4, feedback techniques were studied to improve low-noise amplifier performance. In this chapter, an inductor placed inside the active feedback loop was proposed to improve the bandwidth of the amplifier. It was found that the inductance of this inductor is very important in determining the bandwidth of operation, but its quality factor is not relevant in terms of this and the other parameters. In fact, a low quality factor improves the noise performance of the circuit, and for this reason, a modified miniature 3D inductor was used. As a consequence, this technique improves the area and also the noise performance.

The size of the area is mainly related to the number of inductors employed in the circuit. In Chapter 5 a circuit methodology for designing UWB RF frontends without the use of inductors was described. To validate this design methodology two receiver RF frontends were designed: a traditional inductor-based design and an inductorless design. Both design use a common gate LNA followed by two quadrature Gilbert cell mixers. In the inductorless version, the LNA shunt peaking load is replaced by a resistive load, and a capacitive degeneration is included in the mixer. Simulation results show that the inductorless topology has better linearity and comparable noise figure and uses less power with the same bandwidth. The silicon area for the inductorless LNA and I/Q mixers is roughly 54% less than the traditional inductor-based design, showing an important area saving and improving the circuit portability.

As a summary, Table 5.2 shows the main features of all the circuits developed in this book.

Bibliography

1. B. Razavi, *RF Microelectronics*. Prentice-Hall PTR, 1998.
2. B. Razavi, T. Aytur, F.-R. Yang, R.-H. Yan, H.-C. Kang, C.-C. Hsu, and C.-C. Lee, "A 0.13 μm CMOS UWB transceiver," in *Solid-State Circuits Conference, 2005. Digest of Technical Papers. ISSCC. 2005 IEEE International*, pp. 216–594 Vol. 1, 2005.
3. R. Roovers, D. M. W. Leenaerts, J. Bergervoet, K. S. Harish, R. C. H. van de Beek, G. van der Weide, H. Waite, Y. Zhang, S. Aggarwal, and C. Razzell, "An interference-robust receiver for ultra-wideband radio in SiGe BiCMOS technology," *IEEE Journal of Solid-State Circuits*, vol. 40, no. 12, 2005.
4. B. M. Ballweber, R. Gupta, and D. J. Allstot, "Fully integrated 0.5-5.5 GHz CMOS distributed amplifier," *IEEE Journal of Solid-State Circuits*, vol. 37, pp. 231–239, 2000.
5. T. H. Lee, *The Design of CMOS Radio-Frequency Integrated Circuits*. Cambridge University Press, 1998.
6. O. Medina, J. del Pino, A. Goñi, S. L. Khemchandani, J. García, and A. Hernández, "A method to build-up an integrated inductor library," in *XX design of integrated circuits and systems conference*, pp. 1–5, 2005.
7. A. Goñi, F. J. del Pino, S. L. Khemchandani, J. A. García, B. González, and A. Hernández, "Study of stacked and miniature three-dimensional inductor performance for RF IC design," in *SPIE — The International Society for Optical Engineering's — Microtechnologies for the New Millennium Design (VLSI Circuits and Systems Conference)*, 2007.
8. A. Goñi, J. del Pino, B. González, and A. Hernández, "An analytical model of electric substrate losses for planar spiral inductors on silicon," *IEEE Transaction on Electronic Devices*, vol. 54, pp. 546–553, 2007.
9. J. Aguilera, J. Meléndez, R. Berenguer, J. R. Sendra, A. Hernández, and J. del Pino, ""A novel geometry for circular series connected multi-level inductors for CMOS RF integrated circuits"," *IEEE Transaction on Electron Devices*, vol. 46, pp. 1084–1086, 2006.
10. J. N. Burghartz, K. A. Jenkins, and M. Soyuer, "Multilevel-spiral inductors using VLSI interconnect technology," *IEEE Electron Devices Letter*, vol. 17, pp. 428–430, 1996.

11. M. Geen, R. Green, R. G. Arnold, and J. A. Jenkins, "Miniature multilayer spiral inductors for GAAS MMICs," in *11th Annual GaAs IC Symposium*, pp. 381–394, 1989.

12. A. M. Niknejad, "Analysis, design and optimization of spiral inductors and transformers for si RF ICs," Master's thesis, Univ. California, Berkeley, CA, 1998.

13. A. Zolfaghari, A. Chan, and B. Razavi, "Stacked inductors and transformers in CMOS technology," *IEEE Journal of Solid-State Circuits*, vol. 36, pp. 620–628, 2001.

14. H. Ahn and D. J. Allstot, "A 0.5-8.5 GHz fully differential CMOS distributed amplifier," *IEEE Journal of Solid-State Circuits*, vol. 37, pp. 985–993, 2000.

15. R. Liu, C. Lin, K. Deng, and H. Wang, "Design and analysis of DC-to-14-GHz and 22-GHz CMOS cascode distributed amplifiers," *IEEE Journal of Solid-State Circuits*, vol. 39, pp. 1370–1374, 2004.

16. R. E. Amaya, N. G. Tarr, and C. Plett, "A 27 GHz fully integrated CMOS distributed amplifier using coplanar waveguide," in *IEEE RFIC Symposium*, pp. 193–196, 2004.

17. H.-L. Huang, M.-F. Chou, K.-A. W. Wuenand W., and C.-Y. Chang, "A low power CMOS distributed amplifier," in *IEEE Annual Conference Wireless and Microwave Technology*, pp. 47–50, 2005.

18. X. Guan and C. Nguyen, "Low-power-consumption and high-gain CMOS distributed amplifiers using cascade of inductively coupled common-source gain cells for UWB systems," *IEEE Transaction on Microwave Theory and Techniques*, vol. 54, pp. 3278–3283, 2006.

19. Y. Yun, M. Nishijima, M. Katsuno, H. Ishida, K. Minagawa, T. Nobusada, and T. Tanaka, "A fully integrated broad-band amplifier MMIC employing a novel chip-size package," in *IEEE Transactions on Microwave Theory and Techniques*, pp. 2930–2937, 2002.

20. J. del Pino, S. L. Khemchandani, A. Hernández, J. Sendra, J. García, B. González, and A. Núñez, "The impact of integrated inductors on low noise amplifiers," in *XVIII Design of Integrated Circuits and Systems Conference*, (Ciudad Real), Nov. 2003.

21. R. Wang, M. Lin, C. C. Lin, and C. Yang, "A 1V full-band cascoded UWB LNA with resistive feedback," in *Proceedings of the IEEE 2007 International Workshop on Radio-Frequency Integration Technology*, (Singapore), Dec. 2007.

22. K. Chang-Wan, K. Min-Suk, A. Tuan, K. Hoon-Tae, and L. Sang-Gug, "An ultra-wideband CMOS low noise amplifier for 3-5 GHz UWB system," *IEEE Journal Solid-State Circuits*, vol. 40, pp. 544–547, 2005.

23. A. Medi and N. Won, "A fully integrated low cost packaged CMOS low noise amplifier for the UWB radio," in *Proceedings of the IEEE 2006 Radio and Wireless Symposium*, (San Diego, CA), Jan. 2006.

24. C. S. Chih, W. Ruey-Lue, K. Ming-Lung, and K. Hsiang-Chen, "An integrated CMOS low noise amplifier for 3-5 GHz UWB applications,"

in *Proceedings of the IEEE 2005 Conference on Electron Devices and Solid-State Circuits*, (San Diego, CA), pp. 225–228, Dec. 2005.

25. Y. Huang and T. E. Chen, "A fully-integrated UWB CMOS LNA using network synthesis approach," in *Proceedings of the IEEE 2006 Microwave Symposium*, (San Francisco, CA), Jun. 2006.

26. F. Gong, K. F. Lam, M. Ismail, S.-B. Park, and J. DeGroat, "A 3-5 GHz frequency tunable ultra wideband LNA for OFDM applications," in *52nd IEEE International Midwest Symposium on Circuits and Systems*, (Cancun), Aug. 2009.

27. J. Sendra, J. del Pino, A. Hernández, J. Hernández, J. Aguilera, A. García-Alonso, and A. Núñez, "Integrated inductors modeling and tools for automatic selection and layout generation," in *IEEE International Symposium on Quality in Electronic Design*, (San Jose), 2002.

28. T. Manku, G. Beck, and E. Shin, "A low-voltage design technique for RF integrated circuits," *IEEE Transactions on Currents and Systems II Analog Digit Signal Process*, vol. 39, pp. 1408–1413, 1998.

29. H. Hashemi and A. Hajimiri, "Concurrent multi-band low-noise amplifiers-theory, design, and applications," *Transactions on Microwave Theory and Techniques*, vol. 16, pp. 290–294, 1968.

30. C.-C. Tang, C.-H. Wu, and S.-I. Liu, "Miniature 3-D inductors in standard CMOS process," *IEEE Journal of Solid-State Circuits*, vol. 37, 2002.

31. A. Ismail and A. Abidi, "A 3 to 10 GHz LNA using a wideband LC-ladder matching network," (San Francisco, CA), pp. 382–383, 2004.

32. A. Bevilacqua and A. Niknejad, "An ultra-wideband CMOS LNA for 3.1 to 10.6 GHz wireless receivers," in *ISSCC Digest of Technical Papers*, (San Francisco, CA), pp. 384–385, 2004.

33. S. Shekhar, J. Walling, and D. Allstot, "Bandwidth extension techniques for CMOS amplifiers," *IEEE Journal of Solid-State Circuits*, vol. 41, pp. 2424–2439, 2004.

34. J. Lee and J. Cressler, "A 3–10 GHz SiGe resistive feedback low noise amplifier for UWB applications," in *IEEE Radio Frequency Integrated Circuits (RFIC) Symposium*, (Long Beach, CA), pp. 545–548, 2005.

35. N. Shiramizu, T. Masuda, M. Tanabe, and K. Washio, "A 3–10 GHz bandwidth low-noise and low-power amplifier for full-band UWB communications in 0.25 SiGe BiCMOS technology," in *IEEE Radio Frequency Integrated Circuits Symposium*, (Long Beach, CA), pp. 39–42, 2005.

36. M. Reiha and J. Long, "A 1.2-V reactive-feedback 3.1–10.6 GHz low-noise amplifier in 0.13 μm CMOS," *IEEE Journal of Solid State Circuits*, vol. 42, pp. 1023–1033, 2007.

37. H. Knapp, D. Zoschg, T. Meister, K. Aufinger, S. Boguth, and L.Treitinger, "15 GHz wideband amplifier with 2.8 dB noise figure in SiGe bipolar technology," in *Radio Frequency Integrated Circuits (RFIC) Symposium, 2001. Digest of Papers*, (Phoenix. AZ), pp. 287–290, 2001.

38. Y. Park, C.-H. Lee, J. Cressler, and J. Laskar, "The analysis of UWB SiGe HBT LNA for its noise, linearity, and minimum group delay

variation," *IEEE Transactions on Microwave Theory and Techniques*, vol. 54, pp. 1687–1697, 2006.

39. C.-T. Fu and C.-N. Kuo, "3–11-GHz CMOS UWB LNA using dual feedback for broadband matching," in *Radio Frequency Integrated Circuits Digest*, (San Francisco, CA), pp. 67–70, 2006.

40. J. Kaukovuori, *CMOS Radio Frequency Circuits For Short-Range Direct-Conversion Receivers*. PhD thesis, Helsinki University of Technology, Faculty of Electronics, Communications and Automation, Department of Micro and Nanosciences, 2008.

41. B. Razavi, *Design of Analog CMOS Integrated Circuits*. McGraw-Hill, 2001.

42. L. C-F and L. S-I, "A broadband noise-canceling CMOS LNA for 3.1–10.6-GHz UWB receivers," *IEEE Journal of Solid-State Circuits*, vol. 42, pp. 329–339, 2007.

43. J. del Pino, *Modelado y aplicaciones de inductores integrados en tecnologías de silicio*. PhD thesis, Departamento de Ingenieria Electrónica y Automática, Universidad de Las Palmas de Gran Canaria, 2002.

44. R. Harjani and L. Cai, "1-10GHz inductorless receiver in 0,13μm CMOS," in *IEEE Radio Frequency Integrated Circuits Symposium*, 2009.

45. Q. Li, Y. P. Zhang, and J. Chang, "An inductorless low-noise amplifier with noise cancellation for UWB receiver Front-End," in *IEEE Solid-State Circuits Conference, ASSCC*, pp. 267–270, 2006.

Index